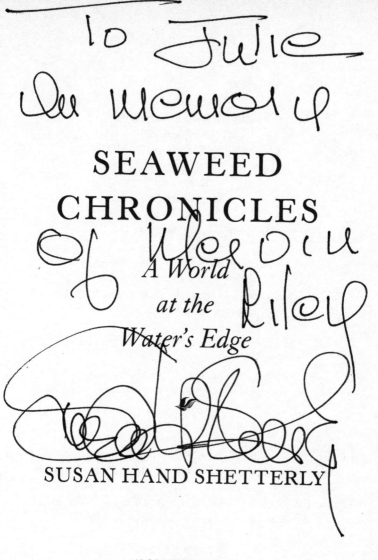

*To Julie*

*In memory*

# SEAWEED
# CHRONICLES

*of Weedou*

*A World*
*at the*
*Water's Edge*

*Riley*

SUSAN HAND SHETTERLY

ALGONQUIN BOOKS
OF CHAPEL HILL
2018

Published by
Algonquin Books of Chapel Hill
Post Office Box 2225
Chapel Hill, North Carolina 27515-2225

a division of
Workman Publishing
225 Varick Street
New York, New York 10014

Library of Congress Cataloging-in-Publication Data

Names: Shetterly, Susan Hand, [date]–author.
Title: Seaweed chronicles : a world at the water's edge / Susan Hand
Shetterly.
Description: First edition. | Chapel Hill, North Carolina :
Algonquin Books of Chapel Hill, 2018. |
Identifiers: LCCN 2018008171 | ISBN 9781616205744 (hardcover : alk. paper)
Subjects: LCSH: Marine algae—Habitat—Maine, Gulf of. | Marine
algae—Harvesting—Maine, Gulf of. | Marine algae as food—Maine.
Classification: LCC QK571.5.M2 S54 2018 | DDC 579.8/3/09741—dc23
LC record available at https://lccn.loc.gov/2018008171

10 9 8 7 6 5 4 3 2 1
First Edition

In memory of my father, Trav, who read me poems
that were read to him as a child, and taught me to love
the sounds and sense of words

To keep every cog and wheel is
the first precaution of intelligent tinkering.

—ALDO LEOPOLD

# CONTENTS

# Seaweed Glossary and Notes

S ome seaweeds in this book are referred to by both their scientific and their common names. In a few cases a species will have more than one common name. To unravel this tangle of nomenclature, what follows is a seaweed list. This is not a compilation of the world's seaweeds, which would fill tomes, but only a list of those referred to directly or by implication in the book.

The word *algae* is often used in the popular press. It can mean either microalgae, which are phytoplankton, or macroalgae, which are seaweeds. Microalgae are one-celled photosynthetic organisms, and those that live in the oceans drift on the water's surface. Macroalgae, or seaweeds, are usually, but not always, anchored to rocks or some other firm surface. They are multicelled.

The word *phycologist* means simply an algae scientist.

## GREEN SEAWEEDS

*Ulva lactuca*: Sea lettuce. Like all green seaweeds, it requires the most sunlight and grows closest to shore. It is eaten by mollusks and crabs and sometimes by fish and people. *Ulva's* distribution is worldwide.

*Caulerpa taxifolia*: The genus *Caulerpa* has a number of species within it. This particular one, *taxifolia,* originally from the Indian Ocean, has become a master invader of the Mediterranean.

## BROWN SEAWEEDS

*Ascophyllum nodosum*: This species of seaweed grows close to shore and provides habitat for many wild lives. It is harvested and made into hundreds of products in enormous quantities for industry. People call it rockweed, knotted wrack, and often—especially in the British Isles—kelp (although it's not scientifically a kelp), or simply wrack.

The *Fucus* seaweeds: This genus contains a number of species that grow along the western North Atlantic, among them *Fucus vesiculosus*, or northern bladder wrack, *Fucus spiralis*, or spiral bladder wrack, and *Fucus distichus*, or flat bladder wrack. They are tough seaweeds and can withstand rough surf. In this book, the genus is referred to as *Fucus* or bladder wrack or rockweed. Both *Ascophyllum* and the *Fucus* seaweeds can be found growing along the rocky shore.

**The Kelps:** Kelps are members of the order Laminariales. They are strong and fast growing. Many species are edible and are harvested for food.

The kelps mentioned in this book that grow on the Pacific coast of North America are the bull kelp, *Nereocystis luetkeana*, and the giant kelp, *Macrocystis pyrifera*.

In the Atlantic, *Alaria esculenta*, or winged kelp, is harvested for food, as is *Saccharina latissima*, or sugar kelp, and sometimes *Laminaria digitata*, or horsetail kelp.

*Undaria pinnatifida* is a Japanese edible brown seaweed similar to the Atlantic seaweed *Alaria*. Its common name is wakame. *Laminaria japonica* is a seaweed in the kelp family used in many Japanese recipes. Its common name is kombu.

**The *Sargassum* seaweeds:** Hijiki is the Japanese name for *Sargassum fusiforme*, a popular edible seaweed in Japan.

*Sargassum natans* and *Sargassum fluitans* are the two free-floating seaweeds that make up the Sargasso Sea.

## RED SEAWEEDS

*Chondrus crispus*, like other red seaweeds, can grow in water up to two hundred feet deep or more, but it is also found in tide pools. The common names for it are Irish moss, on this coast, and carrageenan in Ireland.

*Mastocarpus stellatus* grows in the roughest surf, tightly attached to ledges and boulders. It withstands ice and harbors many small invertebrates on which birds feed.

*Palmaria palmata* grows in the intertidal zone. It thrives at the lowest ebb of the tide. It is called dulse along the western Atlantic, and dillisk in Ireland.

*Porphyra umbilicalis*: This species of seaweed grows along the western Atlantic coast down to Virginia. We call it laver. The genus *Porphyra* grows worldwide. In Japan a number of species of *Porphyra* are raised in large aquaculture projects and are sold as nori.

# SEAWEED CHRONICLES

# PROLOGUE

Downeast Maine, where I live, is for me the most beautiful place on earth, even in February, even on a dark day in a sharp wind. It is ledge and cobble, spruce and white pine, mudflats that glisten like a harbor seal's wet pelt, low-tide rocks covered in layer upon layer of seaweeds, and a horizon straight east across the water into sunrise and Canada. No frills. It has been for me, and I think for so many others who live here, William Blake's grain of sand—a teaching place—and we have learned something of the world from it. Within the wild fabric of this shore, in its many coves and bays, seaweeds and other lives—from barnacles to fish to birds—are bound together, as they are along the shores of other places in the world. It is a tightly woven warp and woof of life, an ancient and essential system of give-and-take.

In April, we can stand at the shore and see long lines of black birds, rising and falling in undulating flight at the water's horizon, homing to their nesting islands. They are double-crested cormorants. They build their nests out of sticks and grass and the seaweeds they've ripped from underwater.

Flocks of robins return. Eastern phoebes come back to the porch eaves. Both, in a cold snap, seek out the windrows of seaweed that lie in the sun above the high-tide line. There, in the warm, rotting tangles, kelp flies and their larvae flourish.

From deep underwater wintering places, adult lobsters lumber toward the inshore waters, where their young find both shelter and food in the rocky seaweed beds. Fish tend toward the shallows, too, to the wealth of food and protection in the dense underwater forests. The common periwinkle moves on its slippery foot onto the rocks in the intertidal zone, where it feeds on microscopic algae, green seaweeds, and the mucus trails of other snails.

Broken strands of seaweed, if they don't wash up in a tide, float in mats out into deeper water, leaky *Kon-Tiki*s carrying a host of small and edible inshore creatures with them before they sink to the bottom. These mats attract fish and birds such as phalaropes, Bonaparte's gulls, and Leach's storm petrels for the sudden gift of food they carry, and often, in migration, birds use them as rest stops. Eventually, all the bits and pieces torn loose in a summer storm, the seaweed detritus, spin into the deep and disintegrate, enriching the planktonic life of the Gulf, and thus enriching all those creatures living in this water, from jellyfish to whales.

But that is not the whole story. A 2016 United Nations report on worldwide seaweed harvests recorded that wild seaweeds cut along coasts in the inshore waters, or as crops in aquaculture farms, had reached twenty-six million wet tons. The countries that harvest the most native wild seaweeds are Chile, China, and Norway. Those top producers of aquaculture-based seaweeds are China, Indonesia, and the Philippines. In Ireland and Iceland and along France's northwest coast, wild seaweed harvesting is a lucrative business, and seaweed farms are increasing on both sides of the Atlantic. The harvest grows worldwide because there are more people to feed, and we find that seaweeds can be made into more things: medicines, cosmetics, soil amendments, edible oils, and bandaging for burns, as well as food and food supplements for us and for our domestic animals. Some have proposed turning swaths of our oceans into industrial kelp-growing farms for biofuels.

In 2011, seaweed-like fossils were unearthed in China. They may be more than six hundred million years old, which makes them among the oldest multicelled organisms ever discovered. It's not surprising, therefore, that seaweed is useful to us and essential to other species. It is ancient and basic, a testament to the tenacious beginnings of life on earth. Why wouldn't seaweeds be a protean life source for the lives that have evolved since? Seaweeds live in a world at the margins, from the high-tide line down to where the sunlight reaches into the water's depths, however dimly. They create a special habitat that seems to be set apart from our world above water. But of course it's not.

THE OLD CHESTNUT that a butterfly flaps its wings and causes a tsunami half a world away is just an old chestnut, but we are learning, most often by breaking complicated and attenuated natural systems, how difficult it is to patch the pieces back together. Sometimes we don't know what all the parts are, and some of those parts we have neglected to save. The years are long gone when we, shore-bound citizens living by the world's oceans, could tell ourselves that our waters would always provide what we asked of them, that supplies were endless, that it didn't matter how much we took or how we did it. Now, faced with the loss of finfish as well as diminished populations of many other species, and with the ballooning effects of climate change and acidification, can we learn from what our oceans are still able to teach us? Because they can still teach us.

Here on this coast—and on other coasts elsewhere—some people are working to restore what they can. To protect what's left, if they can. And to create new jobs for the jobs that have disappeared with the damaged fisheries. Nothing is easy or simple about this balancing act, which stretches from the value of seaweed soil amendments for industrial farms to the buoyant life of a common eider duckling.

As our seaweed harvests in Maine increase faster than we could have imagined just a few years ago, we are asking urgent questions about our lost fisheries and this relatively new business. Have we learned, for instance, how to take—but not too much? And what's too much?

IF YOU ASKED people to guess the next big harvest that we will take from the world's oceans, how many of them would say seaweed?

They might turn over in their minds what they know about the Atlantic cod fishery and its collapse, maybe the American and European eel fisheries and their collapse, or the collapse of the sardine fishery in Northern California and the North Atlantic, or the vanishing of the anchovy fishery off the coast of Chile. They might mention the loss of the wild Atlantic salmon, or the recent implosion of the shrimp fisheries in the Gulf of Maine and the Gulf of Mexico. They'd say: Right. It isn't those.

Nor could it be the Atlantic halibut, or swordfish, yellowtail flounder, abalone, or Alaskan king crab. They, too, are depleted.

What's left? they'd ask.

Most of us cannot get through a day without meeting sea-weeds in a disguised and processed form in toothpaste, pud-ding, pie fillings, and other soft foods, in makeup, soaps, dog and cat foods, cattle feed, and farm fertilizers. Many people in the world, especially in Asia, eat seaweeds daily as vege-tables—sugar kelp, *Alaria* (wakame) and laver (nori), carra-geenan (Irish moss), and dulse, wrapped, stirred, chopped or sprinkled, dried or steamed or simmered in soups.

We can't save what's left of our wild and natural resources, or keep our inshore waters free of pollution and alive with native species of all sorts, without providing people who live by and in these places with jobs that sustain them. Seaweed harvesting supports coastal workers who need the work, breathing a bit of life back into small-town fishing commu-nities gutted by the loss of fish. It offers jobs for biologists who study how wild creatures use inshore habitat, for indus-trial scientists who study what else we can make from it, for the people who pack it and sell it and ship it, and for those who work in factories who make it into the things we use.

*Rockweed* is the word the Maine Department of Marine Resources uses for *Ascophyllum nodosum*. This species is by far the most harvested seaweed in the state, and it is wild cut and used in industry. In 2016, records show that Maine rockweed landings came in at 17,367,229 pounds.

That doesn't include other species that are wild harvested or grown in aquaculture sites. We are new to the pressures of this accelerating seaweed harvest, which means we have less repair work to do than some other cultures that have ransacked their inshore habitats. But we know a lot about plunder. We've been doing it ourselves for years.

THIS BOOK IS about seaweeds and seaweed harvesting. It is also a collection of stories about individual people who work and live at the shore and what they have shared with me of their lives. And it is about wildlife—fish, birds, snails and clams, the tiny scuds, and the big eagle throwing its dark shadow across the bay. They teach me about what's worth saving.

# CHAPTER 1

# The Gulf of Maine

At his studio on Prout's Neck, in 1885, Winslow Homer completed his iconic painting of a Gulf of Maine fisherman, *The Fog Warning*. In 1883, when he was forty-seven years old, Homer had moved to this peninsula, which lies on the east side of the Scarborough River estuary, a few miles south of Massacre Pond, the site of seventeenth-century battles between settlers and the native tribes. The peninsula reaches straight into the Gulf of Maine without any island buffers. From his studio on the second floor of his converted carriage house, the painter began his late, great works of weather and rocks and water—and, of course, the people for whom this was home—that have become a part of the American imagination. In a real, immediate sense, the Gulf of Maine belongs to all of us through these canvases, which tell something of who we are in the world.

You probably know *The Fog Warning*: the fisherman rowing his dory to the mother ship, a dark bank of fog rolling in across the water toward him. Because of the water's swell, the inside of his dory is pitched upward in our direction, and in the hull lie two enormous dead halibut, the beautiful, tasty monster fish that were once common in our inshore waters. By Homer's time, the halibut catch had just started its nosedive, and inshore halibut fishermen hired themselves out to larger ships that sailed offshore for the fish that remained. This is what you see in the painting, a fisherman rowing his catch to the ship, hoping to close the gap before the fog erases all sign of her. He is no longer an independent inshore operator of his own boat. And the fish he's caught are at the end of plenty.

Today the painting shocks us with the wild beauty and formidable danger of our former fisheries, and a warning not of fog but of how quickly a good thing can disappear. We are moving into a time along the Gulf of Maine that forces us to reappraise this once astonishingly rich body of water, and instead of assuming, as we have for many generations, that it will take care of us, we are realizing that we are the caretakers of what is left of it, and that what is left is changing even as we try to understand all the small and large parts and how they fit together. We've worked our way through whales, bluefin tuna, halibut, cod, haddock, herring, lobsters, scallops, soft-shell clams, and bait worms— on and on—and finally to harvestable habitat, and that habitat is seaweed. Some people call this "feeding down the food chain": you harvest the wild creatures the land or water provides, starting with the largest, and end up eating the smallest creatures or, in this case, their habitat.

I believe that to begin to understand just where we are now, we need to know the stories of the bodies of water we live by. We need that sort of grounding. Here by the Gulf of Maine, we who have seen the depths of winter come and go don't forget the impact of snow and ice. They have formed our landscape, our seascape, and, to a degree, our character.

In one of the first poems Americans read by Robert Frost, he tells us how the world might end: "Some say the world will end in fire, / Some say in ice." The Gulf of Maine was made by both. First came the fire. Immense landmasses, sliding on seas of molten rock, slowly drifted together and fused into a supercontinent three hundred million years ago. Eruptions from this violent joining thrust up mountains and buried old seafloors. When the land broke apart, about two hundred million years later, it ripped fiery rifts and opened trenches along its lines of separation. Eventually, what is now North America disengaged from Europe (leaving a slivered edge of Ireland here along the coast of Maine and into Atlantic Canada), and the deep bedrock that spread between them filled with a new ocean: the North Atlantic.

Then, a long pause of thousands and thousands of years as the Laurentide Ice Sheet to the northwest of North America built into a mile-high glacial slab. Like a giant snowplow, it began to grind its way toward the southeast. For most of the Pleistocene epoch, it slid forward, melted back, pushed forward again across rivers and lakes, into valleys, and across mountaintops. Propelled from behind by the buildup of snow settling into layers of ice, it was pulled by gravity, and it carried within it the land it had dislodged. Everything it could break off, it subsumed: water, milky till, rock scrabble and mud flushed from its sides. When it reached the water's edge, it dumped in its load of rock and gravel, sand and mud, building a series of underwater banks and altering the seascape as much as it had the landscape. In its gigantic melting, it raised the sea level and turned what was left of coastal mountains into islands. And that is how the Gulf of Maine was born, with deep troughs from the ancient fire and the high, cobbled hills made by ice.

It reaches from Cape Cod, which is shaped like an arm bending at the elbow—that elbow has been responsible for protecting the cold waters and dense fogs of the Gulf of Maine from the warming waters of the south—to Cape Sable, Nova Scotia. Between them lie thirty-six thousand square miles of what were once some of the most abundant fishing waters in the world, with seventy-five hundred miles of shoreline, including the circumferences of the larger islands. Where it is rockbound, lush forests of seaweed take hold, swaying in the icy water and steep tides. Up in the Bay of Fundy, at the head of the Gulf, the tides can rise to fifty-two feet and drop as far again, leaving fishing boats high and dry.

The Gulf of Maine has the highest and the lowest tides in the world. This means that some species of seaweeds will grow to reach the water's surface at high tide. They create a dense habitat.

The high underwater moraines, built from the piles of glacial rubble, are Stellwagen Bank, Georges Bank, and Browns Bank. The average depth of water above them is about 250 feet, but there are shoals no more than a hundred feet down, and at Cashes Ledge, there are peaks twenty feet or so below the surface. Beside the banks rest the basins—some twelve hundred feet in depth—that were made by that ancient fire, and between Georges and Browns lies the Northeast Channel, a deep entry trench through which flows the Labrador Current.

Such topography, as well as the abundance of islands near the coast, which act like stones in a streambed, sending the water into circular bores, encourages upwelling, bringing to the surface cold, nutrient-rich water that fosters the growth of abundant seaweeds.

Cold water holds more dissolved oxygen than warm, which means more phytoplankton and zooplankton thrive in it, but as the earth warms, changes in circulation patterns result, and there is less mixing in the Gulf of Maine now, less flow. The water used to move up and down and around underwater mountains, and into and out of the dark, cold underwater valleys, as the various currents folded the deep bottom waters into the warmer surface waters, a process that created what biologists call boreal soup, that life-giving planktonic excess. The resulting blooms of life made the Gulf of Maine one of the most valued bodies of water in the world.

Since 2004, these movements have been harder to predict. The nutrient stirring is falling off, and scientists call the change a regime shift. Recent calculations suggest that the water is warming here at a faster pace than in most other bodies of water in the world. Whether it is a persistent trend may no longer be a question, and into this warmer space now come species of fish and crustaceans and tunicates and seaweeds from farther south, life from away.

I RECEIVED IN my inbox an e-mail from Robin Alden, founding executive director of the Maine Center for Coastal Fisheries. Paul Venno had been on the board of the organization for many years, and I had asked Robin if she thought he might be a good person to begin this story about fisheries past and present in the Gulf of Maine, and to link the fisheries to the new interest in harvesting seaweed.

She wrote back this: "Paul's family has farmed and fished in Cape Rosier since before the Revolutionary War. He has a naturalist's eye honed by the love of catching things and his biologist training, and tempered by a sense of place. He's a steward—not just for today but for the future."

And so I begin with Paul. As Robin noted, his family has been a central presence on Cape Rosier on the Blue Hill Peninsula in Downeast Maine since the 1700s. He's compact and muscular, with strong features and a big, gentle presence, and has dug the tidal shores for clams and fished these waters from his boat ever since he can remember. After high school, he left for college, took a degree in forestry, and came back here to live, fishing for shrimp, scallops, ground fish, and lobster, an independent inshore fisherman.

For a number of years, he told me, he worked in federal fisheries, making and implementing policy, and for some more years he worked for the state in what is now the Maine Department of Marine Resources. But he always considered himself a fisherman working with policy makers, and his allegiance was with the fishing community he had grown up with. Yet he knew that both the community and the policy makers were making mistakes. He is seventy-three now and has seen some of the boom and all the bust of the many fisheries in the Gulf of Maine.

Paul started by talking about bycatch, which means catching fish in your net that you are not targeting and that you sometimes can't sell. "When I worked for the Feds, I'd go out to Stellwagen Bank off Boston with the fishermen on their boats to monitor the catches. They were hauling up fish and using these little fish picks to snag and throw the ones back they didn't want. Most of the fish they threw over were dead or dying and it was wasteful. Wasteful. Stellwagen's mostly a sanctuary now. Thank goodness. It's all gravel and sand—moraine stuff—a place where fish can still spawn.

"Another time, I was taking samples of juvenile fish from Gloucester to Eastport, counting the creatures. The Feds were trying to predict the abundance of herring because the herring industry, the people who owned the canneries, wanted to know what the future would bring. We caught very few herring then, lots of other stuff: copepods, elvers. But very few herring.

"If you ask me, I think this is what happened: animals have a place where they hang out—haddock, cod, herring— and they go to the same areas to spawn. If you trawl there, they're spawning, that's what they do. They don't mess with trying to escape the nets. So over the years the trawlers have been to all the places where fish spawn because that's where the fishermen catch the most. And before fishermen were told by the Feds to adjust their nets with a larger mesh and a cutaway top, the mesh was too small, the nets too big, and they caught everything." He pauses, as if seeing the reckless-ness of it all again, and shakes his head. "Fishermen have taken fish at the wrong time and in the wrong way. If you catch fish with a hook rather than a net, you're not going to catch spawning fish because they don't eat when they're spawning. But people wouldn't do that. They had millions of dollars tied up in those trawlers.

"Right now we've got a lot of lobsters and a lot of rock crabs. But the more lobsters they catch, the more careless they are. They throw them around. You've really got to treat them like eggs. They're fragile, especially when they first come up out of the water, and these guys are coming in with two thousand pounds of lobsters. You can't be careful with that many. If they didn't bring in so many, perhaps the price would be better.

"They've got these giant boats—hundred-and-fifty-thou-sand-dollar lobster boats. Or more! They're trying to break even, but they're their own worst enemy. You know, you ask a billionaire, 'How much is enough?' And he answers, 'A lit-tle bit more.'" Paul's laugh is low, ironic. "There are people out there who just have got to have a little bit more."

"There's our avarice," he continues. "Then there's the changing conditions of the ocean. It's warmer. More acidic. I'm not sure we can improve the future or bring the fish back just like they were."

When I ask him what he thinks of seaweed harvesting, he says, "I think seaweed is a good fishery, yes. That stuff will grow back. I used to gather it. I sold crabs and lobsters to an outfit and they needed seaweed a couple of times a week to pack them in. I'd just go down with a clam hoe and cut it at my shore and it came right back.

"But it's all about the machinery you use. If it's a guy out in a dory, and he fills his dory with seaweed, it's fine. You get someone out there filling a giant barge, hundreds and hundreds of pounds a day, that's something else. The small fisherman doesn't do harm. It's the big operations that do."

In 1497, John Cabot sailed for King Henry VII of England, reaching Newfoundland and perhaps sailing south as far as Cape Breton. He probably did not enter the Gulf of Maine, but nonetheless what he found were fish, astonishing numbers of them, the vast majority of which were cod. It wasn't until the beginning of the seventeenth century that reports from expeditions to the New World began to record the sea life in the Gulf of Maine. Again, notice was taken of the size and number of cod. A single fish could weigh up to 120 pounds.

The contemporary Maine historian Edwin A. Churchill writes of the seventeenth-century explorer John Smith, who for a time was a French pirate's prisoner, "During his months of captivity, he occupied himself writing his Description of New-England, a thorough and convincingly laudatory discussion of his 1614 voyage. . . . Published in

1616 along with a map of the region, its success was almost instantaneous. Smith filled his account with glowing stories of the riches to be recovered from the Gulf of Maine—the 'strangest Fish pond I ever saw,' as he called it. Fishermen began sailing to the region, and in less than a decade they had established small fishing stations from the Piscataqua to Damariscove Island.'"

And thus, from such rough-knuckled outposts, began the exploitation and decline of one of the world's best fisheries. This destruction was, of course, not intentional, and it took generations to complete, and it most definitely was dangerous work, but at that time, the wild world seemed enormous, overwhelming, without limit, too big to fail.

WHEN WE SPEAK about the oceans, we don't always picture them as three-dimensional places. Most of us, because we live in this world of land and sky, see in our mind's eye the water's surface and the pitch of the waves, islands at the horizon, and gulls in flight. It is like looking down at a jungle canopy and seeing leaves. The leaves are there, of course, and important, obviously, but that's just the beginning—or maybe the end—of what jungles really are.

I carry around in my mind Cashes Ledge, one hundred miles off the Massachusetts coast in the Gulf of Maine. Conservationists and scientists have fought hard to keep it from the blunt pressure of encroaching fisheries, appealing to people who live by the Gulf to support its continued protections. And because of their efforts, and their photographs, and their love of the place, we are learning to see—feelingly—something of its topography and extravagant beauty.

An underwater camera at Cashes Ledge worn by a diver scans the lush seaweeds and catches the big, heavy-jawed profiles of schooling cod. They're still here. For now. Five hundred square miles around the ledge has been off limits to some, but unfortunately not all, fishing, and those federal protections—although continually challenged—are still in place. If there are jewels in the crown of the Gulf of Maine, and there are many, Cashes Ledge may be the one that concentrates and reflects the most light today.

The ledge is a stab of bedrock that rises off the ocean floor, a small mountain range about twenty-five miles long. One of the richest ecosystems left in the Gulf, it encompasses habitats of mud bottom, mountainside scree, sheer slopes, and projecting shelves. Its highest peak, Ammen Rock, lifts toward the water's surface, and on it, sugar kelps grow thick and very long, creating, with other seaweed species, a startlingly dense forest, one of the deepest tangles of seaweed forests in the world. Schools of cod, pollock, haddock, and cunner prowl through it, and starfish, sponges, sea anemones, and horse mussels cling to the rocks around it. To this bounty come bluefin tuna, blue sharks, right whales, and humpbacks to feed.

Those who are committed to a future in which Cashes Ledge will be permanently preserved speak of how it gives science a window into a time now lost where a whole host of interconnected lives were born, grew, settled, and thrived. Cashes is a portrait of yesterday, but it is also a hope for tomorrow: if we can pass laws that keep it free from human harvest, it may serve as a template for how underwater ecosystems, undisturbed by fishing pressures, can work—especially now within the grip of climate change—and over time it may adapt to the changes and thrive, and serve as a nursery and a model, sending out into the greater Gulf some of its many riches. Those riches will be good for fishermen.

# CHAPTER 2

# The Underwater Forest

*How It Works*

When my children were small, I took them to the shore. It would be low tide, and we walked over the pebbly mud and parted the seaweed strands, the bladder wracks and the knotted wracks attached to the big rocks that the glacier had dragged with it from miles away. We peered beneath the seaweeds. The outer layers had dried in the air, but the under layers held a briny wetness that made the creatures we found within especially bright: starfish; the egg capsules of the dog whelks, small sea snails whose eggs look like tiny Greek amphora; green crabs as new and small as my children's fingernails; young green sea urchins; limpets; sideways-swimming scuds; yellow periwinkles; sometimes a hermit crab or a sea anemone.

It seemed right somehow to be bringing young and growing children to the edge of the bay where life had evolved so far back in time that it was hardly imaginable, as if this place with its seaweeds were the proof we needed that we had come from a world of water, and that everything might have looked, at one time, something like this.

When we are children, our psyches tend to become imprinted on the places we know and love, and for many of us, that edge where water and land meet is one that stays with us all our lives. I didn't think of it then, but now I believe I was offering them exactly this: their home place to imprint upon so that they might go into the larger world with a sense of where they come from, and thus a sense of who they are.

Lifting the seaweeds and finding life beneath them reminded me of the crepe paper balls my sister and I used to find in our Easter baskets when we were young. We would unroll them across the floor of our New York City apartment, where they spilled out treasures wrapped in tissue paper: tin rings with glass stones, and little metal animals that clicked. But these saltwater surprises, sheltering beneath the seaweed at low tide, were better. They were alive. My children and I gazed into a forest, not a tree forest that stretched its branches into the air, but a forest of overlapping, protective fronds, resting like sheaves against the rocks. We were looking for treasures, not noticing then that the real treasure was probably the seaweed itself.

After we left, the tide came back and the various species of seaweeds lifted into the water. The long fronds of the knotted wrack stood at full length in a high tide. The ribbon weeds and the sea lettuce would bend in the water's pulse. The purple laver and the Irish moss would start to glow as the sun reached them through the polish of the rising water. And the lives within them began to stir. Fish moved over them and into them, feeding. The crabs skimmed down from their low-tide hiding places and scuttled across the bottom. The ducks would come, the mergansers hunting fish and crabs, the black ducks puddle-dunking in the shallows for snails and clam worms, and the elegant female eider ducks escorting their buoyant ducklings.

SEAWEEDS ARE ALGAE, a word of Latin origin that once implied a primal ooze, a genesis of original and elemental stuff. They are not plants, in the strict sense. They have no roots, no leaves, no stems. Not really. Only a few species have developed vascular tissues, and these are only sieve tubes that transport nutrients between them, not like the vascular tissues of land plants, with their complex transport of sugars from the leaves down, and water and minerals from the ground up. Essential to land plants is a sophisticated coordination between specialized cells that have distinct jobs to do. When cooperation and distinctions between cells are present in seaweeds, they are much simpler. Each seaweed cell uses the water in which it lives, taking from it the nutrients it needs.

They are called seaweeds, but scientists still can't pin them down with a satisfactory all-encompassing definition because species keep slipping in or out of the categories we have constructed for them. The ancient fossil record is sparse, and little can be done to bring to the fore a clear image of their history.

Sometimes, casting a look back, it is difficult to distinguish a single-celled alga from a bacterium. That's because they both speak to beginnings, when first things—experimental, most of them evanescent—merged and separated and borrowed from one another.

Cyanobacteria were the first living things. They exist today much like their fossil ancestors: they live in water, and manufacture their own food from sunlight. Some species are toxic and dangerous.

They used to be called blue-green algae because of this miraculous ability to turn sunlight, mixed with carbon dioxide and water, into food, which is the prime occupation of algae and land plants. Over time, these photosynthetic creatures floating in prehistoric seas created the biggest revolution on earth. Nothing matches it, not the Chicxulub asteroid plunging into the Yucatán Peninsula and upending the age of dinosaurs, not the wrenching apart of Pangaea, the Paleozoic megacontinent that fractured into the continents we know today.

Minute though they are, cyanobacteria changed the earth's atmosphere by adding one ingredient: oxygen. And here's the marvel: they floated for a few billion years, then were joined on the water's surface by a floating one-celled alga. Two tiny, astonishing beings floating around together for another billion years or so, as the algae and the cyanobacteria took the energy from the sun to make food and dispersed into the air their tiny gifts of oxygen, which, over huge amounts of time, grew into the matrix of all life to follow.

Marine algae, both the one-celled phytoplankton floating in the oceans today and the seaweeds anchored to our shores, and also seaweeds that float free, such as some species of *Sargassum*, supply the atmosphere of the earth with at least half its oxygen, which is the air we breathe. In the process of photosynthesizing, algae also disperse oxygen within the water to aquatic animals, including fish: the air they breathe.

We find one-celled algae in fresh water, on the damp trunks of trees, on wet rocks, on snow, in rain pools, and in the ocean. The phytoplankton that floats on the surface of the Gulf of Maine and all other saltwater bodies, a soup of many different species of one-celled algal life, are called microalgae because they are very small. The seaweeds that rim our shores to form underwater forests, or grow untethered and floating in deeper water, are the macroalgae. They are multicellular and gigantic by comparison.

Because seaweeds look much like land plants, one might assume that the hostas and lilies and such that we grow in our gardens evolved from them, that red, green, and brown seaweeds slowly made their way out of the sea and onto the land. That is not so. Land plants and particular green and red seaweed groups may share a possible evolutionary point of origin, but what we are witnessing in most of them is an example of parallel evolution: land plants and seaweeds came up with similar shapes as the best way to live.

They both need to anchor. Land plants have roots. A seaweed has a holdfast. Some holdfasts are shaped like disks, but others form a fist with many fingers called haptera. Holdfasts attach to rocks, wharf pilings, breakwaters, clamshells, anything rigid and stable. They are made up of thick tissue and fine hairs and a glue-like substance that sticks them in place. Seaweeds that anchor to corals secrete an acid that wears away a little carbonate chink of the coral into which they insert their attachment filaments.

While holdfasts anchor as roots do, they don't transport water or minerals up into the algae they anchor. What they do is set the seaweed in one place and keep it there through tides and currents and storms, as water bathes the porous cells with the nutrients they need. The few seaweed species that grow in a hard sand or clay bottom use a rootlike structure called a rhizoid to penetrate the substrate. These seaweeds with rhizoids are in the green algae group and are probably distantly related to land plants.

The simple parts of seaweeds begin with the stipe, a stalk that lifts up from the holdfast. It looks much like the thin trunk of a young sapling, or the stem of a large grass plant. It takes the seaweed into the light. The blade or frond or thallus (often more than one word in the seaweed lexicon can be used for the same thing) is what the stipe brings to the light. The blade is the equivalent of branches and leaves on a tree. A seaweed blade may branch, and air bladders may punctuate it, especially in seaweeds that are long or that grow in quiet waters where currents may not lift it high enough to the surface and into the light. The bladders are simple buoys carrying the blades aloft. Bathed in water and sunlight, the blades have two important jobs: photosynthesis and reproduction.

Some *Sargassum* species can reproduce by fragmentation, but most species of seaweed reproduce by alternating generations. They have a sporophyte phase, which sends out a drift of tiny spores, as mushrooms and ferns do. And they have a phase of sexual reproduction, as flowers do. With the diploid form of reproduction—the spore phase—a seaweed can replicate itself, but this allows for no genetic variation. Sexual reproduction—the haploid form—introduces the possibility of a genetic mix, but it's extravagant because the microscopic male and female reproductive cells are dispersed into the tide and swept into the enormity of turbulent currents. Most of them never find each other, never join. As a result, the solution for most, but not all, seaweed species is to depend on both means, which allows for genetic change as well as a chance at abundance.

*Ascophyllum nodosum*, or knotted wrack, is the tough, familiar seaweed along our shore. It reproduces only through the haploid form. Its blade is multibranched, shaped like a hardwood tree. At the tips of all these "branches" it begins its period of reproduction by growing receptacles, egg-shaped pouches that mature over the winter into male and female gametes, and when the water warms to the temperature the seaweed requires, it will release eggs and sperm in a cloudy mix into an incoming tide.

If you go to a shore with a marked incline, you will find that the native seaweeds arrange themselves neatly into bands. The green algae live closest to shore, the brown seaweeds inhabit the inshore waters and can also thrive in subtidal depths where sunlight reaches, but it is the red seaweeds that can live in the deepest water with the least light.

~

## The Otter's World

The bull kelp, a brown seaweed of the eastern Pacific, found in inshore waters from Baja California to southeast Alaska, can grow ten inches a day and up to sixty feet in a season. Buoyed by a single fist-size air bladder on a bare stipe, it lifts from its holdfast and spreads wide blades out across the top of the water, maximizing its exposure to sunlight. Another brown seaweed of that same coast, the giant Pacific kelp, is called the sequoia of the sea, a perennial, fast-growing canopy species that can reach 148 feet. Along with bull kelp, it creates forests of unparalleled ecological value. Kept afloat by numerous delicate air bladders, the giant kelp puts forth side blades from the stipe, like the branches and leaves of a tree, creating dense underwater habitat.

Sea otters require this gargantuan forest to survive. There are two subspecies of sea otter, one along the California coast, another along the coasts of British Columbia and Alaska and out to the Aleutian Islands. Sea urchins, the spiny, slow-moving grazers related to starfish, also depend on kelps. And this is how this particular wild system works: Urchins graze kelps near their holdfasts. They also feed on young, tender kelps newly attached to rocks. The sea otters eat the urchins. At the water's surface, the otters float on their backs and socialize and consume what they've caught, draping the long blades of the kelps around their bodies to anchor themselves.

It's a life-sustaining relationship, kept in balance by the otters: the sea urchins feed on the kelps; the otters feed on the urchins, preventing the kelps from being overgrazed and the underwater forests from becoming decimated. As with the old-growth forests on land, these spectacular Pacific kelps have created a habitat that has supported communities of many lives—many species—for thousands of years, protected by the appetites of the otters, the equivalent of swimming Loraxes.

Sea otters, members of the weasel family, grow coats the thickest of any mammal in the world, which led to their near extinction by the early 1900s. When hunting them for their luxurious pelts was finally outlawed, they began to rebound. As they rebounded, the kelp forests, devastated by urchin grazing, grew back. But recently, and at first mysteriously, sea otters have declined again along the northern Pacific coast and out into the Aleutian Islands. Why? It was a scientific puzzle. Not only were their populations dropping, but in some places they had disappeared, especially along the Aleutians, and the Steller sea lions,

A few biologists began to hear reports of orcas, otherwise known as killer whales, hunting the otters. These attacks were made by the transient orcas, a separate grouping from the resident orca populations. The transients migrate up and down the coast from Southern California to southeast Alaska. The reports seemed an exception to the rule, because otters are low in body fat, much too small a prey item for a six- to ten-ton whale, and the information was new, unevaluated, and therefore suspect. But reports of attacks continued, and after a while, scientists connected the loss of baleen whale populations, primarily the gray whale, to the deaths of the otters: these transient orcas had shifted their diet from whales to concentrate more fully on, first, harbor seals, then fur seals, and then the mighty and aggressive Steller sea lions. The sea otters as prey item came last. They were snacks, but the hungry whales needed snacks, the more the better.

Orcas are the largest species of dolphin in the world. They've been called killer whales forever, but that's not quite correct: the transients are whale killers. Unlike other whale and dolphin species that feed primarily on fish, krill, jellyfish, and squid, and unlike the resident coastal populations of orcas that feed primarily on salmon, these transients are more like wolves. They can take down animals much larger than themselves by overcoming them with coordinated pack work. The historic numbers of baleen whales were missing along the coast, and the seal and sea lion populations in the Aleutians were also dropping for a number of reasons. The otters were left, and the hungry orcas turned to them. In some places, the transient orcas cleaned out entire bays. With the loss of otters in the bays, the sea urchins quickly tore away at the kelps.

Jim Estes, one of the scientists who made the connection between the otters and the baleen whales and the orcas, has made a further connection: "What we've seen is that the kelp forest ecosystem in South Western Alaska went from being robust to being . . . gone. It's staggering that it occurred over such a large area in such a short time—just a few years. The food web interconnectivity—that urchin explosions could be linked to whaling 50 years ago—is amazing."

Who knew that sea otters need baleen whales to survive, and that those missing whales had helped keep the great kelp forests alive?

## The Sea within an Ocean

One of the most elaborate wild systems in the world we call the Sargasso Sea. The seaweeds that make up the Sargasso are two species, *Sargassum natans* and *Sargassum fluitans,* both of which can propagate themselves by fragmentation and spend their lives unanchored and floating in the North Atlantic subtropical gyre, a circular system of ocean currents that press in on every side. Although these *Sargassum* species probably originated long ago in the Caribbean, close to shore, with holdfasts, they have adapted themselves to living here now in open water. They can do this because their cells are "totipotent," meaning that any cell within them is capable of making an entire new individual.

This peculiar sea of seaweed is about two million square miles in size, located in the North Atlantic off the coast of Bermuda. An archipelago of drifts and porous islands sometimes twelve feet deep, the Sargasso carries within its lacy fronds ten species found nowhere else in the world, including fish, mollusks, and crustaceans, all adapted to the sea's being neither land nor open water but a bit of both. The current at the sea's western edge is the Gulf Stream, flowing north.

These seaweed islands are also hosts to fish, crabs, worms, and other tiny lives found elsewhere on seaweeds by the shore, but out here they are part of a biome uniquely its own. Not only do young sea turtles seek refuge and food here, but to it come many other species, among them humpback whales, basking sharks, dolphinfish, and bluefin tuna. And this is where the young of the American and European eels can be found freshly hatched.

Mature American eels live in our freshwater lakes and rivers for as long as twenty years. Eventually they begin their journey to salt water. They are females. On their way down they meet the males that live in rivers and streams closer to the coast. As they go, both males and females are transformed, becoming what we call silver eels. They stop eating, their eyes enlarge so that they can see in the dark, the pectoral fins at their sides widen, and a muscular blue-silver gloss shines through the skin of their backs as their bellies turn an eerie white. Just on the lip of winter, in the cold nights of autumn rains, they begin their journey. When they reach deep water, they turn south toward the Sargasso Sea, the only place in the world where they spawn. After they spawn, they die.

We know they do this, and it must be deep beneath the floating seaweed beds, though no one has found a single eel egg stuck in the *Sargassum*, nor a single body of a spent spawner. But odd little fish shaped like tiny elm leaves with pointed snouts have been found in and around the seaweed. They are the larvae of both the American and the European eels that will eventually wend their way into the Gulf Stream to drift north, the American eels dropping out along our eastern seaboard, as the European eels continue, circling below Iceland and then down along Scotland and south.

All along the eastern coast of the United States, they turn inshore in springtime, seeking fresh water, moving between underwater cobbles and through the sheltering seaweeds in our bays, heading toward our streams and rivers as they change first into what we call glass eels, which look like transparent spaghetti pieces, each with a beating red heart and two black dots for eyes, and then into the thin, dark baby eels we call elvers.

Eel populations along the US and Canadian coast and in Europe are near historic lows owing to years of over-fishing, dams that block their access to upstream waters, turbines set into dams that chew them up, and contaminates. Individuals and conservation groups have pressured the US Fish and Wildlife Service to list the American eel as endangered, but there is a problem: managing eels is about as hard as holding on to one because they swim long distances across state and international boundaries. And there is much about them we don't know. Although biologists who study these eels will tell you that they are threatened and, yes, endangered, a federal protection plan for them has not been written.

With over nine hundred licensed elver fishermen, Maine is the only state on the Eastern Seaboard to maintain an elver and glass eel fishery except for South Carolina, which issues a few licenses each year. Canada prohibits catching glass eels and elvers. Europe has closed all selling of young eels outside its territory. Within its territory, it does allow catches for restocking programs.

Those caught here in fyke nets as they come to our freshwater streams and rivers get packed up and shipped to China. There they are fattened in tubs and, when grown, sold to Japan for *unagi*. The Maine fishermen who net them have made spectacular, jaw-dropping windfalls. They are the same people who have been sidelined by the closing of other fisheries because of overharvesting, and they need this work. But one must ask: What's wrong with this picture?

At the Sargasso Sea, boats come to harvest the floating mats of seaweed, and trawlers pull their fishing gear through it, cargo ships regularly traverse it, and within its slow, continuous gyre, plastic trash is collected and held—forever. There is no other sea like the Sargasso, and there are no other eels like the American and European eels. They are bound together and irreplaceable. To work to protect one, we need to extend protection to the other. In 2014, five governments, including that of the United States, came together in Bermuda to see if they could begin to work on conservation for the Sargasso Sea. It's a small start.

For most of us, the Sargasso is a dreamlike place. It is vulnerable precisely because it is far away and such a strange, wild world of its own.

## A Wild Gift

Close by, our pebble beaches are edged with storm-tossed windrows of seaweed, knotted wrack and bladder wrack, that lie in a long, narrow fringe at the tide line. And before you know it, still in the heat of summer, the shorebirds have come back from their nesting grounds in the Far North. Peter Matthiessen called them the wind birds. Swift in flight, the various species funnel down in their migrations to hunt small crustaceans and other creatures that populate the high-tide line within the tossed-ashore seaweeds. These birds are urgent feeders, feasting at shorelines that must supply them quickly with what they need for the huge metabolic task of flying long distances.

Late one afternoon, at the end of summer, my neighbor and I lay on the thin strip of gravel beach in our town, chatting and watching as the bay water, heading for low tide, inched down the shore. It was the end of August, and we'd both stopped work for the day. Just beyond our toes ran a narrow twist of seaweeds, a curled combination of bladder wracks and knotted wracks that the receding water had left behind. Its colors were dull green and mustard and dark brown, and the sunlight gave it a shiny garland look that made me think it would make a lovely Christmas decoration, hung above a doorway, festooned with tiny lights. I was thinking about the feasibility of seaweed garlands when suddenly two birds, as if they were made out of the weed itself, appeared in front of us. They were smaller than the length of my hand, one slighter than the other, and they seemed to pay us no attention as they worked the briny weed, walking daintily ahead and plucking so fast that even though we could have reached and brushed the crisp feathers of their backs with our hands if they had allowed it, we couldn't see what they were eating. They were finding food, but whatever was keeping them going at such a pace was invisible to us.

Perhaps they had been here for some time, checkered birds, the color of rust and black pepper and tortoiseshell. They were so much the color of the wrack that we noticed them only because they moved in quick, short thrusts as they fed.

They were least sandpipers, *Calidris minutilla*, the young of the year, down from the sedge-rimmed bogs of the North, where they nest beside cold-water pools in the tundra and the taiga forests. At most, they are only six inches in length, and a well-fed one weighs about an ounce. These tiniest of all shorebirds appear on our coast both coming and going to summer nesting grounds, to wintering grounds, from the

Ungava Peninsula and Newfoundland to North Carolina, Mexico, and Brazil. They count on these shorelines for food.

It was a moment you can't force—an abrupt epiphany. But epiphanies such as this need an open beach, a drift of damp seaweeds, plentiful critters that live within and around those seaweeds, a withdrawing tide, and a good nesting season so far north that most of us will never visit that soggy, wild landscape where these sandpipers brood their buff-colored, spotted eggs, and where the parents, not long after the young have learned to fly, leave them to the insect-rich pools and head south.

The young before us are on their own, moving down with others of their species, feeding on beaches, in marshes, in the coves between ledges. They know how to find food and to find one another. Gathered in tight, white-flashing flocks in flight, they know how to make their way to places they have never seen before with no one to show them how to get there.

southwest tip, where the corral she built is located. It's on the far side. Out of sight. To the lee of the wind. She says the sheep will be there, too, huddled below it. She adds that the upward slant at the tip of the island, which shields them from the direct cut of the wind, indicates a drop-off onto the ledge below.

Flat Island looks like a lousy place for a flock of sheep. Or anything else, for that matter. But it isn't. And the reason is that the wind whisks away most of the island snow, leaving hummocks of Kentucky bluegrass and Canada bluejoint, introduced species of grass so tough and persistent you'd never want to come upon them in your garden, but heartily welcomed by sheep on a winter island. All the sheep have to do is scrape a hoof to find it, sometimes still a bit green in the blades. And then—the saving grace—there are the kelps. The tides rise and fall on a winter island, and the storms lash against it, bringing the loosened seaweeds that grow in deeper water—the long-bladed kelps the sheep will readily eat—up to the shore, an abundant and clean food.

The relationship between islands and sheep and kelps was cultivated centuries ago in Ireland and Scotland and was brought to this coast as custom by the first white fishermen-farmers. Pasturing sheep on islands (about thirty islands off this coast are named Sheep Island and over forty named Ram Island), and introducing grasses that persevere and that sheep will eat, used to be commonplace. Donna may be one of the few island shepherds left in the state today, but the work has a long tradition. It isn't that she's a throwback, exactly. It's rather that she's taken on the ethic of hard work and physical skill and simple needs that are still part of traditional life here. We could have been standing on the mainland facing this island a hundred years ago, talking about sheep, shouting into the wind. I am guessing that it looks as if little has changed. But looks, as we know, are deceiving.

Turning away from Pleasant Bay, Donna and I get into her gimpy car and drive to her land. She walks ahead of me to the cabin she built and lives in. The path she's shoveled through the new snowfall is precise and thin, and her tall, lanky frame moves in it, one boot set down directly in front of the other as if she were walking a tightrope.

At sixty-one, with steel-colored hair and an open, ready smile, Donna has worked with sheep for thirty-seven years, piloting her outboard runabout to tend flocks on three islands: Big Nash and Nash, where she tends others' sheep, and Flat Island, where she keeps her own flock. Although she grew up around Los Angeles, she's a person who has found a home in the Downeast ways, the sort of life that this coast used to provide for everyone. Its give-and-take is basic and renewable.

Donna shears sheep in season, sets rams out with the ewes a few weeks before Christmastime for breeding, oversees the flock's general health, especially during lambing season, brings the lambs—called saltwater lambs—to the mainland for slaughter in early winter, and sells the meat. When not caring for sheep, she makes rolling pins and wooden bowls and has worked closely with botanists, identifying island plants, including those of Mount Desert Island, where she is listed as one of the authors of the seminal book *The Plants of Acadia National Park*. Donna's life is all work, and work for her is much like serious play. I've been told she can turn a bowl and slaughter a lamb with swift and equal skill, and she takes a practical, tough joy in the world around her and her place in it.

The benefit of islands for sheep, she explains, is that they need no fencing, although they do require some sort of corral for shearing. They don't need buildings to keep hay bales dry, because they don't need hay bales. And if the shepherd chooses good islands, appropriate for sheep, the fleeces of the animals on them grow even and thick and very clean. The islands Donna manages—hers and two of her neighbor's—have acres of open ground, and broad ledges ring them where the sheep gather at low tide to eat the daily flotsam of kelps.

Sitting in her well-ordered cabin, beneath a center post made from a maple bole, she tells me about her flock. They're Romneys, a breed with a dense, long-fibered fleece. Known for their ability to thrive in rain and cold, they were first bred on the Kentish marshlands of England and are the premier sheep of New Zealand, raised for both wool and meat. I had expected her flock to be the semiferal North Ronaldsay sheep, a delicately boned breed kept by the Scots on the northernmost Orkney Islands for their wool. Those sheep feed exclusively on kelps. Over hundreds of years, they've adapted to the salts, iron, and arsenic found naturally in kelps, and their bodies can extract enough copper from seaweeds to keep them healthy. But North Ronaldsays set down on a mainland sward, on a diet of grass that supplies more copper than they are accustomed to, quickly succumb to poisoning, which demonstrates how, over time, species can adapt in vital and complex ways to what their environments can or cannot offer.

Because the North Ronaldsays are dependent on seaweed, their internal clocks are set to the tides, not to sunup and sundown like mainland sheep. Even in the dark on nights with no moon, even in winter, and in a hard wind, the Ronaldsays will clatter onto the ledges to feed at low tide. When the tides are full, they retreat from the shore to ruminate until the water falls away again. Donna's flocks probably do much the same.

In Iceland today, farmers have begun to rely more and more on kelps for sheep fodder. Studies have proved that the Icelandic sheep cannot metabolize enough copper from the seaweeds; pregnant ewes that ate only seaweeds were giving birth to lambs with disorders. However, by supplementing the gestating ewes with copper, Icelandic sheep farmers are seeing their lambs born healthy, and their ewes eating up to twenty pounds of kelp daily and thriving.

We return to the car and drive to a farm that belongs to a friend who owns the flocks she cares for on Nash and Big Nash. They have just rescued a ewe from Big Nash.

"She'd been cast," Donna tells me.

When I ask her what that means, I find, surprisingly, that it does not mean that the ewe was tossed from an abrupt drop, but rather that she rolled over—slipped, perhaps, or was pushed by the wind. We enter a shed through a shawl of bright, blowing snow and find her lying in a bed of straw beside another older sheep chewing its cud.

The cast ewe is cloudy in the left eye. Out on the island she must have been banging her head trying to get up, Donna tells me, and has probably bruised her cornea. She tilts her head at us.

"She's been out there cast for quite a while," Donna says quietly.

When a sheep rolls onto its back, it's like a beetle, but worse: a beetle will eventually figure out how to get one of its six legs to act as a fulcrum. A sheep is stuck. Its legs point to the sky. In the rumen, the section of its four-part digestive system where the essential bacteria and fungi and protozoans—like sourdough starter—keep digestion alive, things begin to deteriorate, and the precious bolus of fermentation starts to drain away.

Donna is lavishing care on this ewe. She has fed her kefir and sauerkraut juice and has gathered kelps at low tide and cut them up and set them in a pan in front of her. She stands over her, reaches down and hoists the ewe's 150-pound bulk, leans to straighten her front legs, heaves her up a little higher and straightens the back ones, then lets her gently down, with the hope that she will, in time, use her legs to stand again.

The ewe's a beauty: pure white, a coat so thick it would be hard to get your fingers through to the skin, a wide, pretty face. It looks as if she's pregnant, Donna tells me. Maybe twins, she says. I watch her coax the ewe to take a bite of the kelp.

I'll be back in spring to go out in Donna's runabout to see the flock and walk the shore on Flat Island. As I drive home, the sun going down, sending a blush through the bare trees and out across the ice and snow of a shallow bay, I reflect on how generations of people have used this land and water to make their lives good.

We have lost the abundance that used to be a part of the definition of this place, and the changing climate bears down especially hard. And yet so many of us work to understand where we live, to use what's left of wild water and shore carefully and well. Perhaps that makes us unrealistic dreamers. Perhaps it makes us steely-eyed realists. Certainly it makes us fighters. We don't want to lose the life here any more than Donna wants to lose her beautiful ewe.

The sun is turning the snow along the rough, ice-tight bay a deeper rose color that flares and fades. It's still bone cold, but the slowly ebbing evening light means spring. Hard to believe that in a month the long strands of knotted wrack, holding tight to the midtide line, disheveled and scoured by ice in the shallows, will be floating in wide-open water, tips swollen, getting ready to release eggs and sperm into the wash of early summer.

IT's LATE MAY when Donna and I head out to Flat Island in her runabout. As we go, Donna skirts a ledge up ahead, barely visible above the incoming tide, then resumes our arrow-straight approach. She tells me that the cast ewe is walking now, back on Nash Island with her flock, and that she bore two lambs this spring.

From a few miles' distance, Flat Island looks like a bit of bog land adrift in the wilds of the North Atlantic, a thatch of shorn green. As we approach, silhouettes of sheep pop up against the horizon. They are big eared, long necked, toothpick legged. Donna and a group of friends sheared them a week ago, and now they look more like goats.

There is something otherworldly about this jot of land, the stark delineations of the sheep, the unbroken horizon beyond it. Other islands fall away from us on either side as we close in, and herring gulls rise in clouds above it, a gray mist of wings and a screech of protest we can hear over the boat's motor. The great black-backed gulls, the bird that Arthur Cleveland Bent called "a king among the gulls, a merciless tyrant," stand over their nests, gorgeous, stark black and white. At last, they rise, too, circle, and float back down. As we approach in the dory we've towed along, a half-dozen Canada geese hike away from the cobble beach.

We pull the dory above the tide line and set out to hike around the circumference of the island. The tide is rising, but there's enough ledge to walk the shore, keeping to the places free of nesting birds, and sheep with lambs.

This place of sheep and gulls and grass and ledge and big sky carries stories you can read, if you know the facts. It's all about history here: layers of the past, the fleeting present, a guess at possible and probable futures. An archaeologist who visited a few years ago, taking core samples from the blanket of dirt, reported to Donna that trees had never grown out here. It's peat and bird scat and seaweeds and whatever else lands on the island, broken down and built up slowly over centuries. The grasses have been given a close mower cut, thanks to the sheep. But by fall, they'll have grown high, almost to our knees, becoming winter fodder for the flock, and down within that thick autumn hay, some new blades will start.

There's no fresh water on the island, but Donna assures me that the sheep get what they need from the grass itself. And when rains fall, which they haven't much this spring, sweet water is held in the bowls and dips of the ledges for a while. It's a thin line between flourish and perish out here. And that seems to me to be an accurate way to see life in this Gulf of Maine—and what all oceans offer, really.

Before Flat became a sheep-grazing island, years and years ago, a colony of Leach's storm petrels probably nested here. They dug their burrows into the old, soft island peat.

In her twenties, when she first came to live on this coast, Donna and her husband, Brad, who died some time ago, were hired by Maine Department of Inland Fisheries and Wildlife to survey a number of islands to find out whether

there were nesting petrels on them. Their instructions were
to step onto an island and smell it. Petrels have a distinct
musk smell, like WD-40 mixed with mink scat. If Donna
and her husband smelled petrel, they were to spend the
night on the island, and in the dark, adult petrels would
come flying in from offshore like bats and crawl down into
their burrows to regurgitate the planktonic food they had
gathered and stored in their foregut for their young.

The delicate birds approach at night to evade the gulls
that would attack them, and they come in singing the strangest warbles, as the babes, underground, sing back to them.
Those are among the songs Donna heard in her first years
on this coast. A chorus of leprechauns.

To me it is astonishing how life shifts: a time of petrels,
and then they are gone and something else replaces them,
and life goes on. By the mid-nineteenth century, herring
and great black-backed gulls were no longer nesting on any
but two offshore islands in the Gulf of Maine, extirpated by
unrestricted egg collecting and the killing of birds to provide feathers for the millinery trade. With state and federal
protections, they rebounded, and then their populations
took off, overwhelming other species such as terns and eider
ducks on the islands. Today, with our open dumps closed;
a diminished supply of clams, sea urchins, and mussels;
and the successful resurgence of bald eagles, which prey on

young gulls, herring gull nests have declined by 17 percent, and great black-backed nests by 30 percent. On Flat Island, during the breeding season of 1995, biologists recorded 468 great black-backed gull nests. Today there couldn't be more than thirty pairs.

As Donna and I walk at the far edge of a small herring gull colony set into boulders and driftwood and dried-up seaweeds and an abundance of jettisoned plastics and storm-tossed lobster traps, the gray-and-white birds rise in deafening protest and settle back quickly as we pass. We spot a common eider's nest, well hidden under a log to one side of the colony, all soft, flecked down and long grasses, with two beautiful gray-green eggs. Wherever the hen is, she's a lonely outlier here.

The sheep and their lambs keep watch from a low hillock as we head south where the land levels out to an extended platter of bedrock, wide and overlaid with cobbles. A tide pool glistens at its far point, to which the geese have retreated at our approach, swimming in nervous circles. Beside us, a wash of cobbles and plastic bottles is heaped over a thatch of grasses. Donna tells me that every year this place receives a new invasion and that it's rising up the island, throwing up rocks of all sizes, and debris, turning pasture into beachhead.

The geese move from the tide pool into the open water, leaving a single gosling at the shore, where it stands upright at the very tip of island rock, watching their retreat. And then we see them: the kelps, some dried out and high on the ledge, others fresh and glowing, emerging out of the sea: sugar kelp, *Alaria*, horsetail kelp, devil's apron. Fresh or dried, they are available to sheep all year, even at high tide.

Donna rips off a piece from a blade of sugar kelp and chews it as we walk. Out here, the picture looks pretty clear: everything else is in flux, but the seaweeds are still a constant in places like this where animals and people come, as they have always done, to make their lives good.

CHAPTER 4

# Ghost Fish

I wasn't thinking of seaweeds as I stood in the broken crusts of snow below the highway in late February two years ago, but they are a part of this story that I had assumed was only about a species of herring and the flow of a freshwater stream.

On that cold day, seven people stood together at the bank of Patten Stream, which runs through my town. Three of us were members of the town's Alewife Committee, and we knew this stream pretty well because we'd been working for four years to get the alewives—silver bellied, each about ten inches long—up into the culvert that stood before us now, its tall cement maw rimmed with ice. The drone of trucks and cars on the road above echoed through its dark chamber, and the stream, about twenty-five feet across, poured through it like a silver tongue and dropped down and ran beside us, tumbling against the rocks and ice in the streambed. A hundred yards to the south, whitecaps were busy whipping up the surface of Patten Bay.

From Labrador to the Carolinas, alewives once returned in uncountable numbers from the ocean in springtime to enter rivers and streams, using them as highways to reach their natal freshwater ponds and lakes to spawn. Here in my town, those ponds are Upper and Lower Patten, about four miles north of the culvert. The fish were a bountiful food for other wild species, from bears to eagles, and for people, too.

As dams and hung culverts and other blockages along their historic range prohibited their yearly migrations, and pollution runoff sickened them, and the remnant populations were overfished, the magnificent schools dropped away. Our run at Patten used to be prolific. But that was years ago. Now only a few fish are able to climb up the water pounding through the culvert built onto the old ledge that was once part of a gentle streambed. You can trace the way the stream flowed naturally in an arc to the east, descending toward the bay in gradual riffles before the culvert was built, forcing it into this rushed and straightened channel.

As a committee, we'd tried building fish ladders out of wood and aluminum, lowering them down with a crane, the boom hovering over the avalanching water of a spring melt. But the fish didn't take to our ladders. We also dipped with plastic-coated nets and lifted the fish into the culvert, holding the nets steady against the current until they righted themselves and swam out and up the stream. The man who owned the property on both banks didn't care for our ladder designs or our dipping, and we knew that we had to come up with something he'd accept. We'd met with hydrologists, migratory fishery specialists, and people from other towns who had struggled for years before they succeeded in getting their alewife runs past all sorts of human-imposed impediments, from town fathers to derelict dams.

But that February day two years ago felt different. We stood with a representative from the National Oceanic and Atmospheric Administration, a coastal habitat restoration specialist, a fishway architect, and the state fisheries biologist in charge of fish that swim from salt to fresh, and from fresh to salt. It was, I felt, an auspicious moment that had come at the end of a long winter, at the end of a long series of failed attempts, an off-season moment because the fish were nowhere in sight.

Claire Enterline, the state biologist in charge of alewives, was going over the reason that so many of the fish—the numbers are in the thousands—come back to this stream despite their inability to breach the culvert: her department has stocked the ponds with adult alewives culled from another river. After spawning in the ponds, those adults swim down the stream, which is a lot easier than swimming up, and for those of us who watch them go, they seem to vanish into the Gulf of Maine, where they school with other alewives in deep water, traveling northeast across the continental shelf and then veering south.

Their young hatch and grow into little fish in the ponds. We call them flippers because they leap into the air at what seems the slightest provocation, the only time in their lives that alewives jump. Eventually these young make their way downstream in late summer and early fall, thousands of them, to spend their first winter in the salt bays around us, in those habitats that include abundant, sheltering forests of seaweeds where they find food and protection—and there they grow up. When they join the adult schools, they are gone for three or four years until, as sexually mature fish, they come back to where they began. Right here—to this town. To this stream.

Claire is a petite, athletic woman with a friendly, matter-of-fact manner, and she explained that her department was not going to stock the ponds again because it was time to give the fish accessible passage here at the stream instead of artificially creating a run dependent on the yearly stocking of the ponds. Our committee, no matter how dedicated, was not moving enough fish to sustain a healthy run, which she put at about 250,000 fish a year. So, she was telling us, we've got to find a way to build a passage into the culvert that the alewives will use and that the owner of the stream banks (and legally he owns the bottom of the stream as well) will accept. In short, it's the task ahead. It may, she said, take us time.

I keyed in on the word "us." I hoped it meant we were now a team. It wasn't that we in town didn't know what the problem was, but finding a solution had been fraught. Standing at the stream that day, I thought that the people had arrived who might finally help us, and that we all wanted the same thing: to build back a wild system. One that includes alewives.

As we stood talking in the snow, the adult fish were swimming somewhere miles beyond us in swift, glittering schools, and I remembered that when the stream reaches fifty-one degrees Fahrenheit in May, the mature alewives start their run, their blue-gray backs the color of the tumbling water. You can stand on either bank and see their dorsal fins piercing the surface all the way down to the bay.

TED AMES IS a person who has thought a lot about alewives and what they once meant to the life of this coast and its fisheries. He was born into a fishing family on the island of Vinalhaven in Penobscot Bay and won a MacArthur Fellowship for his work on Atlantic cod in 2005, work that turned out to be as much about fisheries history as about science, because cod, once prolific (the fish that built a nation), have mostly disappeared from the Gulf of Maine. The term that is used to describe them is "commercially extinct."

As a nation, our environmental history includes species that are gone forever: the passenger pigeons that once flew in migrating flocks so large they blocked the sun, the Labrador duck, the sea mink, the great auk, and on and on. Shadows at the door. It is said that schools of cod once stretched from bays outward fifteen miles or more into deep water. Like the pigeons and the ducks, they seemed limitless. How could a bird like a passenger pigeon disappear? How could a fish like a cod disappear? But the passenger pigeon is gone forever. And today, with the lowest numbers ever recorded in the Gulf of Maine—the cod fishery has declined 90 percent since 1982, 77 percent in the past five years—cod are at 4 percent of what a managed stock should be, and the commercial fishery is closed.

Ted says, "We caught the big ones, and we caught the little ones. And then it collapsed." Where they endure in force is almost exclusively in the memories of long-retired fishermen, or in the records of fishermen long dead who meticulously wrote down their catches. It's hard not to think of Ames at work as a Holmes-like investigator, pursuing the smallest clue back to its source, weaving together a passel of disparate cod stories and catches of the past, constructing a schema of old spawning territories, and searching to find some order in all this absence. He was fishing for cod, but they were ghosts now: big predator fish that used to arrive inshore in vast numbers hunting just about everything that crawls or swims, including crabs and lobsters, squid, young haddock, Atlantic herring, and, perhaps especially, alewives.

Why would someone take on such a task as this? The answer is to revive a wild species and a human culture, and a way of life for both, and to know, at last and in detail, how the life of a specific species of fish worked within this body of water, what went wrong, and exactly what the cod would need in order to thrive again. The cod is not a fancy or a beautiful fish, but not too long ago we made it the lodestar of our coastal communities. It built livelihoods for some, great wealth for others. Gutted and cleaned, its flesh is a luminous white; cooked, it flakes into pure, thick chunks. Here in New England, traditional recipes for cod are many—and good—and in England, Scotland, Spain, Norway, Iceland, and France, it is the same. Some of the older recipes call for salt cod, or *bacalao*, as they say in Spain. It kept well because the salt preserved the flesh before the days of refrigeration.

I buy fresh cod once in a long while at the market in town. The last time I asked our local fishmonger where the fillets lying on ice before him came from, he told me they came from the Pacific: *Gadus macrocephalus*. We were standing a quarter of a mile from salt water where our cod, *Gadus morhua*, once swam, abundant, irreplaceable.

Today I am sitting across from Ted on the top floor of a two-story building at Stonington Harbor. This is where he and his wife, Robin Alden, founded the Maine Center for Coastal Fisheries after he won the MacArthur Fellowship, and where they work to preserve a future for coastal fishermen.

Ted has a compelling presence, his face roughened by the years out in the glare of the sun off the water, his eyes clear, with a direct, self-contained gaze. He tells the story of his grandfather's warning: if you go into fishing, you've got to get another trade, because fishing is too unreliable and you're too small.

He enrolled at the University of Maine at Orono, earned a graduate degree in biochemistry, and then came home and fished for twenty-eight years, learning how to manage and conserve his own strength handlining, gillnetting, dragging for scallops, trawling, and lobstering. The degree in biochemistry taught him how to think like a scientist, and his life out on the water taught him almost everything else. Like the fisherman-scientist Paul Venno, he straddles two worlds, and over time, as he saw that he and his fellow fishermen were taking too many fish, he grew interested in policy.

"Way back in the nineties I began this work," he says, "because I saw that management was not required to take anecdotal or non-peer-reviewed information into their analyses, even though the fish stocks were collapsing, and that nothing was being done to find out about the fine-scale peculiarities of cod in the Gulf of Maine. So I started a spawning-ground study. I felt that some research might help solve the problem we had with the policies that lacked a scientific basis for a solution."

Fishermen had not been asked by the policy makers to contribute because the policy makers assumed that they already had all the information they needed, that fishermen were by nature storytellers, and that cod ranged the Gulf erratically without much cohesion or fidelity to territory. But Ted was aware that what the fishermen knew was detailed and narrative—and accurate—and that it might hold untapped promise. From them he found out that cod were still abundant in the 1920s, and after conducting interviews and studying the old logs, he teased out the fact that there were four distinct, self-reproducing groups of cod in the northern Gulf of Maine that returned to their specific inshore spawning grounds year after year. In the fall, some cod would spawn offshore, but in the spring, others came to the mouths of streams and rivers to spawn. He discovered that there was an order to their movements, and there was specificity to the groups that made up the whole.

"Each of the four subpopulations of cod in New England moved along their own migration corridor, pursuing the seasonal migration of prey such as Atlantic herring and alewives," Ted explains. "I began doing work at Muscongus Bay because the towns have alewife records that go back to the nineteen twenties. I started by using historical alewife landings in conjunction with the distribution patterns of young-of-the-year alewives shortly after they leave their natal rivers.

"I circled the areas where they were located in the fall and early winter, when cod normally are migrating offshore, and then overlaid my historical cod study results on top of it, and discovered the grounds with ripe cod were located inside the circle of young alewives. So I plotted out the movement patterns of cod and other predator species such as haddock, pollock, and white hake and found that in the fall, while many of them were migrating away, many others were migrating into the alewife circles—just like the cod. The research showed the same point: the big fish stayed inshore as long as the prey was abundant. And the prey near Muscongus Bay were juvenile alewives."

It is the loss of these forage stocks, Ted believes, as well as the years of overfishing, that caused the collapse of the fishery.

The juvenile alewives living in the inshore bays for their first year of life counted on seaweed forests. So did Atlantic herring. So did young cod. Subtidal communities with submerged vegetation, such as eelgrass and a variety of seaweed species, create habitat for these vulnerable young fish in their first year. Studies found an abundance not only of young cod but of cunner, hake, lumpfish, tomcod, and winter flounder in these inshore habitats.

"As fish grow they need larger shelters, so they shift from one shelter size to the next, then to the next," Ted explains. "If those are widely spaced, they get eaten. They need to be able to get from one shelter area to another. Seaweeds can help them do that. Rocky areas, sandy areas, kelps and other seaweeds: fish need a plethora of habitat types—we call them benthic structures—rich, diverse habitats that allow these critters to grow.

"Eventually, once fish like cod grow and reach a point where they are too big to hide in these coastal waters, they end up schooling and start preying in these same habitats," Ted tells me. "You're either hiding from something or you're hunting for something that's hiding. Seaweeds form one of the initial areas of protection for them and then become a feeding ground. If you disrupt this, you eliminate a life stage. Seaweed's a niche that plays an important role for fish. If you clean it out, then you're back to square one, because you're missing a key part of a complex ecology that photosynthesizes, provides shelter, and modifies wave impact. A whole suite of critters has evolved to take advantage of its presence."

What Ted has taught us from his years of paying close attention is a story about where we live: Once, aggregates of cod followed the herring and alewives north, and they used these inshore waters in spring to spawn. In the fall, groups of cod gathered inshore again to spawn and to feed on the bonanza of young alewives that were migrating down the rivers and into the bays, where they remained through the winter. When the dense alewife populations thinned and the Atlantic herring all but disappeared, there wasn't much left for the cod to eat.

If we want a fighting chance to return a robust population of cod to their ancestral waters in the Gulf of Maine, we need the Atlantic herring and the alewives, among other forage fish, to back them up. That means protecting the herring from an incautious fishery, and it means removing culverts and dams for the alewives or providing them with state-of-the-art fish ladders. We do this for the alewives themselves, and for the dream of returning cod.

As Arctic ice sheets melt and waters emerge out of the lost ice fields, concentrations of cod are shifting north to the water between Iceland and Norway. But what about here in the Gulf of Maine? Without an abundance of prey such as alewives and herring, can we expect them ever to return in force to a body of water that is warming? I put that question to Ted.

"I know that improving alewife and Atlantic herring stocks will create conditions along the New England coast that should enhance the recovery of the coastal stocks of cod and haddock and other predator fish species," he says.

"My sense is we fail to appreciate how resilient marine and aquatic systems are once all the pieces are together. Given half a chance, I expect the cod will rebound. Ocean warming is occurring, but they should do fine around here for the next century—at least—if we bring back the fish they eat. But I guess we'll just have to wait and see."

Cod is a fish held in the minds of people who still dream of its historic abundance. This dream of cod restoration comes with a cat's cradle, stringing together the repairs and protections of an entire ecosystem. If we take on this somewhat quixotic task, as we are beginning to do (a job that requires hard work, a lot of money, and good science), it isn't worth much unless our lakes and rivers that run to the Gulf are clean enough to support fish that migrate into fresh water—alewives and smelt and bluebacks and shad, sea-run brook trout and eels—and unless our bays are clean enough so that seaweed beds, eelgrass beds, cobble, sand, and mud bottoms can create their complex topography of place. Sometimes when a wild population falters, it becomes mysteriously fragile and unresponsive to efforts to revive it, a peculiar and troubling phenomenon called the Allee effect. And yet, science indicates that fish stocks are probably not, in fact, subject to this effect. They can diminish, and if protected, as Ted indicated, they do bounce back.

A case in point is the alewife fishery in Maine's Kennebec River. Since the Edwards Dam and other impediments were removed from the Kennebec in 1999, alewives have quickly reestablished their runs. Now the alewives come in as prolifically as they did in the past before the dam. In 2014, more than two million of them swam up the Kennebec into the Sebasticook and to Benton Falls. And that doesn't count the alewives harvested below the falls, or those fish that veered off into freshwater ponds downstream. The Kennebec now hosts the largest alewife run on the East Coast in recent history.

As a result, a few cod, shadowy forms in the high tide, are ghosting the mouth of the big river.

IT'S THE END of a dry October, with brief hard rains in the last week raising the level of the streams. The young-of-the-year alewives are letting the flow of the water take them from their natal ponds into the streams and down to the saltwater bays. At Patten Stream you can see them if you wait long enough: here, a small dark back, its caudal fin propelling it forward, and over there a silver flash as a little fish turns sideways, and sometimes a sudden splash as a fish jumps.

Our brand-new pool-and-weir fishway was finished two days ago, just on the lip of the downstream juvenile fish run. The owner of the land on either side is putting in a gravel road down to it. Now the construction workers, who sandbagged the stream to move slabs of gray granite as big as hay bales into place with an excavator and shovel, have gone.

They had cut a path down to the stream through brush and trees and piled the granite slabs in a corner of that upturned earth until they were ready to set them. The granite came from the abutments of an old bridge up-country, the one my kids, when they were young, called the singing bridge because cars going over its metal span made it hum. Now, set into this streambed in semicircles, the curves directed upstream, each with a central notch, these slabs of the singing bridge have been turned into five weirs, each one higher than the one before it. They have transformed the stream into a series of rising steps. Over them the water tumbles, creating not a hum but a baritone rumble.

Everything is new. The road. The path. The granite lying on top of the ancient schist in the water looks beautiful, but so new. Give it time. The water will carry down gravel and sand and small twigs and leaves—all the flotsam of a streambed—and they will lodge in the open cracks and crannies of the weirs and seal them. Eventually, freshwater algae and lichens will find purchase on the stones. They'll darken it and give it pattern and texture. Time will make this new thing beautifully old.

This is the end to more than six years of work. Next May, the adult alewives will return, and we'll be waiting to see how they like the new riverbed, and whether it raises the water high enough to create the resting pools they need, and moderates the pitch of the spring melt enough, and allows them to reach and swim through the culvert.

For now, the young fish, born from the eggs of the adults we caught in nets this past spring and carried up the downward-rushing water and released into the culvert, are doing something both natural and miraculous: they are changing from freshwater fish to saltwater fish. I am sitting on one of the granite slabs as a few of them slip by. Promise and danger await them. A brave new world.

There is nothing like watching thousands of fish fight their way upstream to their spawning ponds in the spring. Except, I imagine, watching caribou herds migrate across the Alaskan tundra. Or watching the sun setting the horizon to the east on fire, as I saw when I stood on a bank of the Platte River in Nebraska at dawn watching the sandhill cranes on their way north. The birds had stopped to sleep on sandbars midriver, one leg up into their belly down. When they awake, their voices bugle into that fierce light and their wings thunder—a gorgeous excitement. Or the monarch butterflies I saw, high up on a Mexican mountain, flitting through the ribbons of mist in the tallest firs and bunching together on branch tips, hundreds of them packed close, like huge, shivering fruits, everything silent except for an occasional breeze through the branches.

Here, of course, the sound in spring is not silence, but hard rainwater and the last of the ice melt flushing out of the woods and overflowing the streambed. But the fish are as silent as the monarchs: silver-sided fish, dark blue-green backs all crowded together, all facing upstream, their fins piercing the pitch of the water.

"We aren't sure how climate change is going to play out," Claire Enterline tells me. "There are major processes that will be altered: ocean pH, water temperature, ocean current patterns, sea level, and freshwater input. How this will impact species, we can only guess based on what we know now.

"But," she says, "a lot does ride on how stable these species' populations are to begin with. That's where our restoration work is important. We may never see the Gulf of Maine like it was before European settlement, there may never be cod populations like there once were, but by doing what we can to minimize the negative effects on what is still around us, we can at least hope to get back to a functioning ecosystem."

A functioning ecosystem is a living mosaic as strong as we can make it.

# CHAPTER 5

# Seaweed Stories

*The Old Ways*

Occasionally you can still find them out on islands, crumbling near the water's edge, the old eighteenth- and nineteenth-century kilns built out of stones gathered from the shore. People on the Irish and Scottish coasts and in Brittany cut and burned seaweeds in the pits of those kilns to make potash and pearl ash, valuable potassium salts.

The wet seaweeds—*Ascophyllum*, *Fucus*, and the kelps—had to be lugged up from the shore, carefully turned and dried, and then burned at a temperature that would render them into products that were sold to make glass and soap, to bleach linens, to encourage bread to rise, and to use as fertilizer to sweeten fields. In the boom time, around 1809, Ireland was exporting about 5,410 tons of potash a year. It was backbreaking work that whole neighborhoods engaged in, and at its height, the many kiln fires created smoke so thick it endangered the lives of nearby pasturing cows. It wasn't long before the seaweeds in some places were overcut, the shores laid bare.

Then, as suddenly as it had appeared, the market vanished when potassium salt deposits were discovered underground in Germany and in Chile, and mines were opened.

The burning of seaweed resurfaced with the discovery that the ash residue could be used to extract iodine. But that, too, disappeared when deposits of iodine were found belowground. Left alone, seaweeds regrew, with farmers coming to the shore to harvest them for their gardens, and gatherers cutting favorite species to eat and to feed to their domestic animals. Over time, the old kilns were disassembled by wind and rain and snow.

My GREAT-GRANDFATHER WAS born in County Mayo, a land of blanket bogs and clay on the western shore of Ireland, facing the North Atlantic. He was just a boy when he sailed to America with his parents in the 1860s. Somehow, they had survived the famine.

Even today, you can see the ghost of the famine that provoked their flight and that of so many others in the ridges on the Mayo hills: old shapes of potato gardens, suddenly abandoned, as if time had stopped. And in a sense, it had. Either the villagers who cultivated the plots were too weak to dig into them anymore, or the digging would have unearthed only blight—or both—and they left them where they were.

Before the famine, the Irish farmers of Mayo and the Aran Islands and other sparse places near the coast, where the ground was poor and the rain was hard, had learned to fashion gardens that produced abundant, life-giving potatoes, enough to feed a growing family and the family cow. They called them lazy beds, and some farming coastal people in Ireland today have initiated the practice again. But a lazy bed is not for the lazy.

If you want to try one, here's how to start: A farmer marks out the shape of a rectangular bed with twine across an open, mown field or a grassy yard, three feet wide, about ten feet long, with a further width of eighteen inches of uncultivated ground along each length. Then she collects seaweeds—kelps, knotted wracks, bladder wracks—that have been storm-tossed into windrows (not living seaweeds, but those that have become unmoored and are piled at the upper shore) and hoses them off to be sure they are salt-free, or she lets the rain gently cleanse them.

She spreads the seaweed generously over the grass within the rectangle at a depth of an inch or two and sets potato seed (chunks of quartered potato, cut and air dried, each of which contains at least one pale, fingerlike eye) into the seaweed that lies against the turf.

Then she takes up a sharp spade (in Ireland they call this a loy) and slices into one of the eighteen-inch borders of turf outside the twine, cutting away from the bed, carving a long strip of grass and root and soil, loosening it so that it is hinged only at the edge of the bed where the twine is. She lifts up the twine and folds that narrow strip over the bed, covering potatoes snugly beneath it. Over and over she cuts strips, turning them onto the bed as if she were turning the pages of a book, or folding sheets, working down one length of the rectangle, then up the other. Thus all the potatoes are covered, top and bottom, as they rest in a nest of seaweed. Only the upended grass roots and dirt show at the top of the bed now.

The twine is taken down, the eighteen-inch trenches are dug out, and the loose soil is mounded onto the bed as the potato plants sprout and grow. Underground, the grass and the seaweeds decompose, nourishing the potatoes, and the trenches on either side get deeper, supplying extra soil when needed and draining off excess rain, which is a problem in Ireland. The beds can be refreshed yearly with seaweed dug into them after the fall harvest.

On the Aran Islands, there is only an occasional thin skin of turf over bare rock to make a lazy bed. The old-time farmers would collect sand from the beach coves, mix it with decomposing seaweeds and what little there was of turf, and make their own soil, into which they set the potato seed, then shoveled more of the dirt and sand and seaweed mixture over the seed, building up the beds with channels for rain runoff on either side.

In his 1907 book *The Aran Islands*, J. M. Synge wrote this: "The other day the men of this house made a new field. There was a slight bank of earth under the wall of the yard, and another in the corner of the cabbage garden. The old man and his eldest son dug out the clay, with the care of men working in a gold-mine . . . for transport to a flat rock in a sheltered corner of their holding, where it was mixed with sand and seaweed and spread out in a layer upon the stone."

It may be a seaside farmer's prejudice, but it's claimed that nothing tastes quite so good as a potato grown in seaweed gathered from a nearby shore.

### Carrageenan

When I was a child, my family spent summers on Cape Cod, where my mother had worked in the theater before she married, and one of our pleasures was for my sister and me to walk the long sand beach hunting for clumps of Irish moss that had floated away from their holdfasts and arrived storm-tossed above the low-tide line. We were searching for the cream-colored, sun-bleached blades. After we had gathered enough, we brought them to our mother in the kitchen of the rented house and rinsed them over and over in cold water, both to dislodge sand and to rid them of some of that whiff of iodine and beach rot.

Then we soaked the clean, slick blades in warm water, tossed the water out, and set them in a pot with milk and lemon zest and a touch of vanilla. We put the pot on the stove, at a low heat, and watched as our mother stirred. The mixture began to thicken. Soon it was coating the spoon. My sister and I found this change magical.

Our mother cracked some eggs, beat the yolks with sugar in a bowl, and poured the thick seaweed mix through a strainer and into the yolks and sugar. She whipped the egg whites, folded them in, and served us custard.

Blancmange, she called it. It was a beautiful off-white linen color with just the slightest taste of the sea.

THE LATIN NAME for Irish moss is *Chondrus crispus*, and in Ireland the common name is carrageenan, an Irish word that means "place of the small rocks," where, in the low and subtidal zones, it's picked fresh. A species of red seaweed, it's a beautiful purple color when growing. It looks a bit like the petals of a dark chrysanthemum.

The substance in Irish moss that thickened the blancmange is also called carrageenan. Science has appropriated the word to mean the emulsifier found in a number of species of red seaweeds. It is a phycocolloid (a seaweed gel), chemically inert, but with an adaptability that makes it ideal for combining ingredients into many products.

Three kinds of phycocolloids figure in the daily lives of people who live in the industrialized countries. They are used for thousands, perhaps millions, of processes. Some uses overlap between them, but in general, carrageenan (from the red seaweeds) is added to certain foods, such as some yogurts, soymilk, coconut milk, almond milk, some tofu, and cottage cheese, and toiletries such as shampoos, hair rinses, toothpastes, body creams—the list goes on.

Agar is another phycocolloid, also derived from red seaweeds and also used in foods. But it is, most famously, an invaluable suspension material for laboratory tissue cultures, a growth medium in petri dishes, and the gel used to separate strands of DNA.

Alginate, the third phycocolloid, is derived from brown seaweeds. It is an important ingredient in wound dressings (some wounds need to be kept moist and covered with a material that is nonabrasive), textile printing paste, coatings for high-end paper, and the fluids used in fracking.

SNOWDRIFTS OUTSIDE MY house frame two photographs I've taped to the study window. They are of a woman up to her waist in water, walking in her seaweed garden in Bali, encased in a warmth I can only imagine. She wears a woven grass hat against the sun. The sky above her is broad and blue, and her seaweed garden grows on either side of her beyond loose stick fences, and it looks from here, from this season of snow and ice, as if life could go on forever just like this: one small bay and a family growing and harvesting seaweeds within it.

Today it occurs to me that she is growing *Eucheuma*, a tropical seaweed eaten in Indonesia and the Philippines and on many islands in the Indian Ocean and western Pacific. *Eucheuma* is also grown and harvested by local people to sell to international companies because high-grade carrageenan can be extracted from it.

FMC, a company with a multinational reach, has at one time or another made everything from seaweed products to pesticides to amphibious landing vehicles. It is the only company in this country that produces food-grade carrageenan, and its seaweed-processing plant sits at the Rockland, Maine, waterfront, where it imports *Eucheuma* from the Pacific.

International trade creates some odd turns and twists: In our waters, *Chondrus crispus*, or Irish moss, grows abundantly, and it was once harvested and processed at this Rockland facility. But today, ships come in loaded with cargo containers packed with *Eucheuma*. Irish moss is rich in carrageenan, though not as rich as *Eucheuma*. The superior quality of the Pacific seaweed is why container ships tied up at the Rockland docks cast long shadows across the beds of Irish moss growing beneath them.

IN NOVEMBER 2016, the National Organics Standards Board of the United States voted to remove carrageenan from use in organic food products. This sets in place a directive that, if it holds, will require all foods with organic certification to be free of any trace of carrageenan. The job will take time, but already some of the largest organic food manufacturers are adapting their ingredients.

The board chose to remove carrageenan from the organics list as a response to studies that have linked it to gastrointestinal inflammation, colon cancer in laboratory animals, and diabetes—an ominous list.

Dr. Andrew Weil, father of integrative medicine, suggests that as a result of these studies, carrageenan should be eliminated from one's diet as a precaution, although he notes that both the United Nations and the World Health Organization have concluded from a number of countervailing studies that carrageenan can, in fact, be safely used in foods, including infant formula. But to make things more confusing, the European Union has decided to ban carrageenan from infant formula, although some studies suggest that carrageenan actually improves gut health. In Ireland, during the Great Potato Famine of the nineteenth century, Irish moss was an abundant food for starving people who lived by the shore. Very simply, it saved lives.

International companies that process carrageenan, such as FMC in Rockland, have stepped into the fight to keep their products relevant and profitable. It may be years before the public knows whether carrageenan is a helpful food additive or a big mistake. Data suggest that the carrageenan itself may not be the culprit; rather, the way the carrageenan is processed, or the contaminated waters in which it's grown in some unregulated places in the world, may make it dangerous.

CHARLESVILLE, A SEASIDE town on the southwest toe of Nova Scotia, Canada, faces into the Bay of Fundy. It is here on a large coastal site that Acadian Seaplants Limited has established one of its most spectacular and most secretive enterprises. The company is growing *Chondrus crispus* on land. The huge facility contains a network of tanks filled with seawater and is in perpetual lockdown as if it were the Pentagon. What it makes are not secret plans for war, however; rather, it replicates the life cycle and creates beautiful forms of the Irish moss originally gathered from coves and bays throughout the seaweed's range. This assemblage of tanks, refreshed with water from the nearby ocean, and laboratories with the highest technology, speak to the ambition and range of this innovative Canadian company.

Tended within the facility by scientists who work to give it a rainbow of color morphs and what's known as "mouth feel," *Chondrus crispus* is meticulously prepared for the Asian market. It becomes a dazzling salad mix, or a blade slipped by a bartender into a dry martini, where it floats like a translucent flower. This is upscale designer seaweed, and so far, it has not been affected by the controversy in the United States and Europe.

## The Mother of Nori

Kathleen Drew-Baker never visited Japan, and yet her studies revolutionized the harvest and consumption of seaweed in that country.

She was born in England at the dawn of the twentieth century, attended the University of Manchester, and studied seaweeds along the northern coast of Wales, back when coal was king there.

In the days before the collapse of the coal industry, the Welsh mines employed whole towns. The miners walked to work carrying lunch pails and came home covered in coal dust. What moves me—always—in the old photographs are the smiles, the miners' faces dark with the mine dust, their teeth white and their smiles real as they face the camera.

The northern coast of Wales is a wild area still, of cliffs and ledges, narrow roads, and shore paths from one village to another. The wind and surf are constant. Out over the water in their summer abundance fly gannets and cormorants, Arctic terns and Manx shearwaters, and from the cliff-side heathlands blooming in wildflowers rise the voices of stonechats, the thrush-like bird whose call is loud and sharp, like two stones tapped briskly together.

Kathleen Drew-Baker came to this shore by train through mining towns, down from the gritty city of Manchester, and collected seaweed specimens to take back to her lab. A smart, dedicated woman, Kathleen became an accomplished field botanist and research scientist. She received the highest honors for her studies at the University of Manchester and was hired to teach. But she lived in a time when academic women, especially women in science, weren't taken seriously if they chose to marry. When she married, she was let go from her job, and in the years before her reinstatement as an honorary research fellow, she worked out of a laboratory built for her by her husband. She also became the cofounder and first president of the British Phycological Society, a group of scientists interested in the study of algae.

Kathleen's specific interest was in cryptogamic botany, the study of spore-producing organisms, which includes many species of seaweeds, and her specialty was the genus *Porphyra*, what we call laver and the Japanese call nori, the delicate, translucent red seaweeds found along a number of cold-water shores around the world, including those in Wales, the western North Atlantic, and Japan. (Only recently, *Porphyra* has undergone a reconfiguring and renaming by seaweed specialists, who have broken it into eight new genera, distinguishable only through DNA samples. I call it by the nomenclature that Kathleen used.)

In the Gulf of Maine, there's a species of seaweed that's easier to find than *Porphyra* at low tide. It's called *Mastocarpus stellatus*, and we often see it in winter. It grows thickly up rocks in heavy surf in short, knobby orange-red rugs, and when the tide pulls back, it's exposed. The common name for it is false Irish moss. Nearby these *Mastocarpus* beds, which are favored by overwintering purple sandpipers,

the other, the shell-inhabiting phase, was the diploid phase. They were, in fact, two parts of the same organism.

Her discoveries eventually had an enormous impact on seaweed farming in Japan, where at least two species of *Porphyra*, or nori, are harvested for wrapping sushi, covering rice balls, and seasoning soup. This beautiful and nutritious seaweed had been highly prized by the Japanese for over a thousand years, and long ago, peasants were allowed to pay their taxes to the emperor in nori—it was as valuable as cash.

Seaweed farmers in Japan traditionally ate fish and oysters from the bays as well as seaweeds, but harvesting *Porphyra* was a strange business for them because the seaweed utterly vanished in summer. They couldn't find it anywhere. By early fall, nori farmers would set out hundreds of bamboo poles in local bays, and sometimes they would string nets from the poles, and—miraculously, as if the poles and the nets had summoned a reclusive nori god—minute germlings came, catching on the nets and poles, and then, as the weather cooled, a thallus or blade would begin to grow on them, very small at first, waving in the water, exposed and then covered by tides, delicate, veil-like soft purple plumes. The farmers would cut the blades a couple of times throughout the winter, carefully leaving unharmed the holdfasts, which are shaped like tiny buttons. As in our own gardens on land, where certain plants, such as chives and parsley, grow back if the roots and stalks are left, another crop of nori would emerge from the holdfast after the first was cut. But the seaweed was mysterious and the harvests were unreliable. Some years there were only a few blades to cut. *Porphyra* seemed to come and go without any way to predict it or to encourage it to stay. Farmers called it gambler's grass.

Though members of the British Phycological Society immediately recognized the significance of Kathleen's discovery as a brand-new piece of seaweed science, it wasn't until much later that phycologists in Japan read her work and understood that she had solved the mystery that had bedeviled them for more than a thousand years: they needed a supply of shells to bolster the nori harvest.

Kathleen died young, in 1957, not knowing that her research along the wild Welsh coast would eventually transform lives in Japan, far away. Her paper had come out in the chaos after the war. Most of the fishermen and seaweed harvesters of Japan had left their work to become soldiers, and the oyster harvesting had fallen away during that time. That resulted in less shell debris in the bays, which meant fewer nori plants. But that wasn't all: a series of typhoons had flooded the bays, washing fertilizers and pesticides into them from the rice fields. It was a period of hunger and disarray.

By the time the Japanese put Drew-Baker's work to use, they had little nori left. Since then, the industry has grown so fast that it now mirrors the soybean fields and cornfields of Nebraska: miles of nori nets and poles stretch along the coastlines of Japan and also along bays in China. In many of these places, there are still prominent village coopera- tives where neighbors work together, using sophisticated mechanization as well as hand labor for all aspects of the growing and harvesting of nori and the making of nori sheets, the most valuable seaweed product in the world for human food.

There is no marker that indicates Kathleen Drew-Baker walked the northern Welsh coast, but its raw beauty pre- vails, and you can walk it yourself and see pretty much what she did over eighty years ago. On a promontory along the coast of southern Japan, however, there sits a small shrine dedicated to Kathleen, where she is honored for her work every year.

Her son, Dr. John Rendle, visited Japan to see this shrine.

"My mother never knew what a difference her work would make," he said. "She studied the seaweed for purely scientific purposes, but without her work the entire indus- try could have disappeared. I don't know what my mother would make of it all now. She'd be surprised that what she did back then has now turned into a multibillion-dollar industry. But I don't think she'd like sushi—she wasn't very adventurous when it came to food."

When she did her field research in Wales, at the height of the mining days, a miner's breakfast consisted of a serving of cockles—the small, round clams dug from the sands at the shore—a rasher of bacon, and laver bread or a serving of laver cakes. Perhaps she never tried it. She might not have trusted it any more than she would have trusted sushi, but it has to be one of the heartiest breakfasts, once beloved by miners, and a part of cultural history throughout the country. Today not many people eat laver bread or laver cakes anymore, although there are traditional cooks who swear by them.

For those who have an adventuresome palate and access to a cold-water shore, here is how the Welsh make the bread and the cakes. But first, it's important to note that in a coal miner's house, the kitchen stove, fueled by coal, would burn all day long, heating the home and providing a place to cook. I mention this because laver requires a lot of cooking.

Laver, or *Porphyra*, grows where there are beds of empty shells of bivalves such as cockles, oysters, soft-shell clams, and mussels. Here on the Maine coast, the coves where laver predominates are mostly found along cobbled Downeast shores, where the tides pull away and the mussel beds and mussel detritus and the lavers are exposed together. Out of water, the seaweed looks like a pale purple scrunch of Saran wrap lying in the sun.

Laver should be gathered without disturbing the hold-fast, then washed in multiple changes of water to dislodge sand and to rinse away some of the salt. Cooks set it in a pot at the back of the stove and simmer it for six hours with adequate water. It turns from a delicate purple to a bold dark green, and the blades disintegrate to a pudding-like consistency. Miners didn't add a bit of lemon juice and olive oil before serving, but modern cooks recommend it.

It is spooned onto hot buttered toast and eaten.

Laver cakes are made by combining the pudding mixture, which can be purchased in glass jars in specialty stores, with oat flour and oatmeal to make a stiff dough that is pressed into patties, then fried in bacon fat. The cakes are served with bacon and a side of steamed or fried cockles.

# CHAPTER 6

# Harvests and History

*What We Know and Don't*

I'm not a seaweed eater—at least that's what I maintain. But it's not precisely true. I consume seaweed every day in products that contain it, along with other ingredients, as do most people in the developed world. When I say I'm not a seaweed eater, I mean I don't cook with it, and yet, writing this book has brought me to the world of culinary seaweeds, and I've eaten what others have prepared, and they've been good. Very good. But are they good for you?

Studies have shown that there remains much we don't know about seaweed and human health. It is, in fact, a field of inquiry that is wide open and that will attract committed phycologists for years to come. We do know that extracts from seaweeds used in finfish aquaculture and in land-based agriculture contribute to human nutrition—a step away from eating seaweeds directly. Dr. Susan Brawley, a research phycologist who teaches at the University of Maine at Orono, contributed to the latest study of seaweed as food, which states, "It can be concluded that knowledge of the beneficial effects of algae and their extracts as food additives for humans lags far behind that on which diets have been formulated for commercially important species in aquaculture and agriculture."

In other words, there's more to do.

Among bits and pieces of many other foods, a small amount of evidence of seaweeds has been found in human middens thousands of years old, but despite our ability to plumb many of the secrets of our past, we have not figured out how to measure the effects of eating seaweeds today. Part of the problem comes from our accelerated technological ability to examine human nutrition and digestion with a scientific refinement we've never had before. Because we can apply advanced technology to understanding the foods we eat, we are reexamining what we thought we knew in everything, not just seaweeds.

Seaweeds do contain surprisingly high quantities and qualities of nutrients—vitamins and minerals, and amino acids, the building blocks of protein, and phytochemicals. These healthy components are in there, but the question is, can our bodies extract and use them? The dense dietary fibers found in seaweeds make it difficult for most people to digest them adequately to access the nutrition they contain. An exception is the Japanese people, whose food culture depends on the sea. They have developed, over centuries, gut flora that can break down and utilize seaweeds far better than people in most other cultures.

On the other hand, the fibers that act as roughage may aid in cleansing the human digestive tract for everyone and contribute to the health of intestinal flora. For instance, the alginates found in brown seaweeds have the capacity to absorb and remove toxins in the human gut and can be used in weight-control programs because they provide a sense of fullness.

One micronutrient that all humans can access in sea-weeds is iodine. This is both a blessing and a caution, because a little is essential, but too much may cause harm. What's too much? Japanese iodine intake from edible seaweeds is among the highest in the world, and a recent study esti-mated that the daily average intake of iodine per person in Japan was probably between 1,000 and 3,000 micrograms, whereas in this country the recommended daily allow-ance for an adult is about 150 to 200 micrograms. Foods other than seaweeds that supply iodine are some fish, milk products, grains, vegetables grown in iodine-rich soils, and iodized salt. Interestingly, milk's iodine content may come from the cleansing of the teats of the cows with an iodine solution before milking as much as or more than from the summer pasture grasses they consume.

Japanese living in Japan are among the healthiest people in the world, despite their high iodine intake. They eat a lot of fish, a lot of seaweed, tofu, and vegetables in the cabbage family, and somehow this mix gives them a better chance at optimal heath than people in many other cultures have. But again, it may be the combination of these foods that's salutary. For instance, both tofu, made from soybeans, and vegetables in the cabbage family restrict the uptake of iodine, and this probably buffers the high-seaweed diet.

Hijiki, the popular Japanese seaweed, has naturally elevated levels of arsenic, as does, to a lesser degree, *Ascophyllum nodosum*, the seaweed common on both sides of the Atlantic (as does rice grown in the southern part of the United States on fields that once produced cotton). Arsenic is a heavy metal that can cause acute illness and death, but seaweeds render inorganic arsenic, the most dangerous form, into organic arsenic, which is less toxic.

Many species of seaweeds are excellent collectors and concentrators of toxic metals and other pollutants, and they are routinely used in ports and bays and river mouths to improve water quality. This process is called phycoremediation, or seaweed fixing. Seaweeds render a number of toxins into less harmful forms as they assimilate them, but those used in phycoremediation are, of course, inedible. Some of the same species are harvested for use in food and in farm fertilizers, and because of this overlap, phycoremediation needs to be strictly monitored.

✔

## When Things Go Wrong

In early August 2009, along a broad stretch of holiday beach at Saint-Michel-en-Grève in Brittany, on the northwest coast of France, *Ulva lactuca*, the common green seaweed of the North Atlantic, lay in deep windrows on the shore. Over the decomposing mass a hot crust had formed, and beneath the crust was stirring a fierce poison.

When a horse and rider broke through, slipped, and went down, the horse died of pulmonary edema within minutes. The rider was saved by a man with a bulldozer who drove into the muck to scoop him out, unconscious. The poison was hydrogen sulphide, a quick-acting gas that builds up as a byproduct of anaerobic decomposition. Inadvertently, local farms had leaked nitrogen-based fertilizers into the bay through runoff from summer rains and from pig and cattle pens, and perhaps a water treatment plant built too close to the shore had overflowed. The seaweed proliferated with this nitrogen-rich diet, growing in deep layers on the beach, cooking its poison in the August heat.

The scene was terrifying. The sudden chaos. The thrashing of the dying animal. Before this accident finally drew the attention of the press, a man who was hired by the town to drive a truck piled high with this seaweed debris, in an effort to clean up the beach, died, as did two dogs playing at the edge of the shore.

The potential for this sort of tragedy exists in coastal communities around the world: a fast-growing seaweed in the shallows of a warming shore, highly responsive to unchecked nitrogen runoff, becomes weaponized.

WHAT KEEPERS OF aquarium fish want in their tanks is a green habitat that isn't eaten by the captive fish or fouled by recirculating water, one that perseveres, no matter the season, the intensity of light, the length of the day. Plastic, yes, a choice. Or you might have tried *Caulerpa taxifolia*. It looks like an underwater fern, but it is a seaweed, native to the Indian Ocean. Many people love aquarium fish, and that includes biologists at the Oceanographic Museum of Monaco, who built habitats in large tanks with *Caulerpa taxifolia* a couple of decades ago and in the process released bits of the blades into drainage systems, which spilled them into the Mediterranean, where they regenerated and thrived, adapting to water colder than that of their native range. And then they began to spread. They grew much closer together than the original form, up to fourteen thousand blades in eleven square feet, and because of changes in the species' behavior, phycologists began to refer to it as the aquarium strain. Swaths of the inshore Mediterranean waters quickly became sites of *Caulerpa* monocultures, outpacing native sea grasses and seaweeds.

*Caulerpa taxifolia* is an especially bright green, a bit neony, and it looks vaguely toxic, which in fact it is. It is eaten by very few Mediterranean sea animals, with the exception of the native salema porgy. Reputedly, when these fish feed on the seaweeds at their most poisonous (summer and early fall), the fish cause hallucinogenic reactions in people who cook and eat them.

However, those who work to keep the busy Mediterranean bays clean of sewage, farm runoff, and heavy metals, which accumulate in the waters around places where high numbers of people live and work, see a virtue in this form of *Caulerpa*. It is a volunteer seaweed for inshore phycoremediation. True, it outcompetes many natives, but it also rigorously takes in and retains in its tissues a range of pollutants, and water becomes clearer under its watch. It is too late to extricate *Caulerpa* from this beautiful, deeply historic body of water. The job has become an attempt to control its spread, through laws against possessing it, and cleanup where new patches begin.

In the United States, under the Noxious Weed Act, it is illegal to sell or transport this aquarium strain. California successfully eradicated a small colony of *Caulerpa* near San Diego by covering it with tarps, anchoring and sealing the perimeters of the tarps with sandbags, and pumping chlorine into the space beneath them. The chlorine killed everything it touched, but the tarps contained it, and the invasive seaweed was stopped.

THERE ARE OVER three hundred species of *Sargassum*, many of them a challenge to identify even for phycologists. Most grow in the tropics, anchored by a holdfast to a hard surface—shell, coral, or rock.

When these species detach from their holdfasts, through storm or age, and the currents and tides are just right to gather the loose blades together, turning them into floating mats, they can foul marinas, catch in boat propellers, and clog fishing nets. The British attempted to bulldoze, handcut, rake, trawl, and suction an invasive species of *Sargassum* that had formed mats along their coast, but the seaweed came back. When an herbicide was freely applied to a bay in Ireland that had been invaded, the *Sargassum* died, but almost everything else in the bay died with it.

And yet it is in the Caribbean and the Gulf of Mexico that a combination of *Sargassum* species has become much publicized and hotly debated. The species that are causing a problem start out growing from holdfasts in the Caribbean, in the Gulf of Mexico, and along the coast of Brazil where the Amazon empties into the Atlantic. They grow old or get pulled loose in the surf and wash ashore on West Indian islands, along the Yucatán, and in other resort areas. Piled high on beaches, rotting in the sun, and extending out for yards into the water, where they move like slush within the waves, the seaweeds meet and greet arriving tourists on whom these countries have come to depend.

Hawksbills are among the rarest sea turtles in the world, with only about twenty-two thousand nesting females in the Atlantic, Pacific, and Indian Oceans combined. Here in the western Atlantic, they nest on beaches in the Caribbean and along the Yucatán and the islands in the Gulf of Mexico. When those beaches are choked with seaweeds, the young cannot dig through the slippery mass after hatching, nor can they scramble over the tangle to the water.

Mats of floating *Sargassum* in the open water provide food and cover for hawksbills and capture heat needed by the very young turtles, but the pileup on beaches kills them.

What causes this plethora that turns a good thing bad? A great amount of attention—carefully scientific as well as carelessly hysterical—is being paid to the problem for a couple of reasons: the first is that people from waitresses to hotel owners to diving operators need to make a living in these places, and the second is that these same seaweed species have contributed enormously, creating a richness in the water fauna in the same way that healthy corals and mangroves and sea grass beds do.

All river water eventually pours into the oceans, and the Amazon discharges more water than any other river in the world. Runoff in the Amazon forests from cutting old-growth trees, contamination from the expansion of large agricultural plots to raise cattle and soy crops, and sewage from the new towns sprouting up on its banks, along with a prolonged rainy season that has always brought the forest to the river and thus to the sea, have altered what the Amazon carries from its inland drainage basin of 2,722,000 square miles.

As currents in the Atlantic carry that dark seam of river water north, the nitrogen and carbon in the overfed water affect the ocean. One of the first noticeable changes is increased seaweed growth. Add to this the pollution dumped into the Gulf of Mexico and the Caribbean, and the warming temperatures of climate change, and conditions are ripe for a sudden, overwhelming bloom.

A long, many-branched living *Sargassum* blade, anchored firmly by its holdfast close to shore and swaying in the water, can provide food and shelter for as many as three thousand lives, tiny lives to be sure, but necessary building blocks for larger ones. A blade like this can also host other seaweeds that attach themselves to it.

In a normal year, detached *Sargassum* either sinks to the bottom, turning the carbon it contains into food for creatures that prowl the ocean floor, or washes onto beaches in modest amounts and stabilizes the sands, building up the dunes and protecting the shore from storm damage. And when hungry migrating shorebirds arrive along these beaches in late summer and early fall, some amount of *Sargassum* at the tide line brings them the food they need. This is how it's supposed to work.

⚘

## An International Reach

D r. Raul Ugarte works for the biggest seaweed harvesting, processing, and research company in the world: Acadian Seaplants Limited of Dartmouth, Nova Scotia. The company was founded in 1981 and handles thousands of tons of seaweed annually, making products with it or selling it to companies that add it to everything from beauty aids to plant-growth regulators. The company is also the inventor and producer of the high-end multicolored *Chondrus crispus*, or carrageenan, in its specialty aquaculture labs on the coast of Nova Scotia. Its products are exported to seventy countries worldwide. If there's a Goliath in the seaweed business, this is it.

Ugarte is a native of Chile, but he has worked for Acadian Seaplants as a research scientist since 1995. Living in Atlantic Canada for many years, he commands a perfect English that is difficult to understand because it's dressed in a heavy Chilean accent and inflection. I have watched at public meetings how native English speakers lean in and fix their eyes on him with a special intensity when he talks because they don't want to lose the gist of his thought, and this linguistic opacity probably confers an added gravitas to what he has to say.

As the lead research scientist for his company, he is a smart, charming phycologist who knows as much about worldwide wild seaweed harvest as anyone anywhere. He is primarily an applied scientist, which means that his job is to make seaweeds available for things that people can use, whereas the goal of a basic scientist is to understand how something works. Actually, I believe Ugarte would say he's a bit of both.

In the past ten years, Acadian Seaplants has extended its reach into Maine to harvest *Ascophyllum nodosum* with catalytic effect: it has forced this state, its coastal citizens, its inshore scientists, both applied and basic, its relatively modest seaweed companies, and the seaweed harvesters who sell to them, as well as bird biologists and fisheries biologists, to grapple with the issue of wild-cut seaweeds and what the future of the harvests means here. Without Acadian Seaplants, we might have gone on for some time in a relaxed, somewhat disorderly, but companionable way.

Now the Department of Marine Resources is prepared to make rules statewide for the harvest of *Ascophyllum nodosum*, the seaweed that is the most heavily cut, at over seventeen million wet pounds a year. It will become the first species of seaweed to be regulated. And one of the scientists at the table, helping to make this policy, is Raul Ugarte.

The company first initiated a harvest in Cobscook Bay in Maine, close to the Canadian border, and it outraged so many local citizens that a plan was written up by the department restricting the company's reach into places around the bay deemed of special concern. Now, Acadian Seaplants moves slowly down the coast, getting closer to the center of Maine-owned businesses with their own cutting regimes. Some decidedly negative press has shadowed it here, as well as over on the northwest coast of Ireland. A few years ago, when Acadian Seaplants purchased a prominent Irish seaweed company that harvests *Ascophyllum* and

eastern Atlantic, is being subjected to the same sort of mechanized destruction. The answer is somewhat unclear.

*Ascophyllum nodosum*, growing from Europe's northwestern coast down to Portugal and from Greenland to the coast of North America down to New Jersey, covers our high intertidal zone along this rocky coast, where it clings with its holdfasts and swings its long blades in the tides and currents. If you travel the shores of the Gulf of Maine in the state of Maine, you know it, because in every sheltered and mildly turbulent cove or bay it presides as a thick, protective fringe of life. But every bay and cove is slightly different. Each offers nutrients and anchoring sites for seaweeds that are unique to themselves, and because of these distinctions it is impossible to say with accuracy just how long it will take a harvested seaweed bed to recover: sometimes *Ascophyllum* grows back very quickly, and sometimes it takes years.

*Ascophyllum* is processed for alginates, added to the food of farm animals, and used in some health foods for human nutrition, but it is as a soil amendment and conditioner that it shows an almost miraculous promise. According to the research from sources compiled by Michael D. Guiry, a phycologist and former director of the Ryan Institute in Ireland, *Ascophyllum* amendments improve soil structure and prevent erosion. Seaweed extracts, made primarily from *Ascophyllum* and used in farming and nursery projects, increase the resistance of plants to frost and increase their uptake of organic matter from the soil. They also improve a plant's resistance to stress, reduce the incidence of fungal and insect attack, facilitate seed generation, and protect plants from damage when used directly on roots as a dip in transplanting.

New Hampshire currently permits no commercial rockweed harvest, and Massachusetts Division of Marine Fisheries doesn't regulate a commercial rockweed harvest because it assumes there is none. *Ascophyllum* beds and the beds of *Fucus* species are left alone in these states for now, except for the occasional citizen gathering modest amounts for garden compost or a lobster bake. In Nova Scotia and New Brunswick, Canada, there are robust commercial harvests of wild rockweeds, as there are in Maine.

Where these harvests occur, you can approach *Ascophyllum* from the water or the shore, and only two things restrict your take: the tools you use and the regulations you follow. A third might be how you understand your relationship to the place where you work and live.

In Maine there are three means of cutting. The simplest is with a knife or machete. It is the low-tech, small-harvest option. The advantage is that the harvester can see exactly how high up the *Ascophyllum* blade he or she cuts. The disadvantage is that you are stepping on beds of the seaweed as you cut. Maine state regulations now require a sixteen-inch leaving, which gives this seaweed, living for as long as twenty years (a few phycologists say much longer, even suggesting that a holdfast might persist for up to a century), a chance to come back.

The second method of harvesting is with a cutter rake, and this is what Acadian Seaplants gives to the workers it hires. It provides each of them with a flat-bottomed boat and a rake with a knifelike blade soldered behind the tines and a built-in dam on either side to prevent too close a cut. The harvester plunges the rake directly down into the seaweed bed and heaves it upward with both hands, slicing the seaweed fronds, then swings the rake over the gunwale and into the boat, shakes the seaweed out into the hull, and plunges the rake back into the water again.

It does a job on the rotator cuffs, but it can get pretty close to a leave limit of sixteen inches as the boat drifts along the shore, unless the harvester attaches a rope to the rake, throws it out into the water, and yanks it, ripping the seaweeds and their holdfasts off the rocks. People standing on land watching some of the harvesters work the tide have documented with photographs this rope maneuver. It's not how they were trained, but it's faster, easier, and kinder to the rotator cuffs.

The third method, used primarily in Midcoast Maine, is the mechanical harvester: a flat-bottomed boat with a rotary cutting head and a suction hose that pulls water over the cut weed and spews it into netted bags. This manner of harvesting has undergone some improvement over the years; mechanical harvesters now use biodegradable fluids, in case of a malfunction, and built-in guards have been installed to minimize bycatch. The bagged seaweed is floated on the water until the tide begins to change, when the harvester picks them up, tows them behind the boat, and heads toward the unloading dock and the boom cranes and the trucks.

A good harvester, with rake or mechanized boat, can cut five tons of *Ascophyllum* a day and can earn $1,500 a week. More rapacious harvesting methods were tried years ago in Nova Scotia and are now outlawed, and in Europe machines are sometimes used to clear-cut sections of seaweed beds that require long fallow periods of regrowth and recovery.

There have been a handful of small but thriving private *Ascophyllum* businesses on this coast for almost forty years. With the arrival of Acadian Seaplants and its potential to overshadow them, they have organized themselves into a group of seaweed business owners, harvesters, and scientists—including a member recently retired from Acadian Seaplants—to form the Maine Seaweed Council. They have given themselves the job of educating the public about what they do and advocating for the protection of their resource and their investment.

# CHAPTER 7

# Starting from Scratch

In the late sixties and early seventies, young people began moving to the small towns of Maine, and many of them came to the towns along the coast. These people were well educated and from urban or suburban backgrounds. What they were looking for wasn't always the same, but they had a shared history. They had come of age in the post–World War II culture of American exceptionalism, which was starting to unravel under the pressures of the Vietnam War, political assassinations, and the brutality of the fight for civil rights. They wanted a new start—a theme that runs through so much of American cultural history. These country towns in Maine allowed them the space, and the land was cheap.

What they wanted may have varied from one to the next, but those who stayed here, who didn't give up, who acclimated to the isolation, the long, dark winters, the lack of money, and the small-town culture and politics, learned to work hard and to listen to the people in the communities they had come to. Others tried it and left.

Linnette and Shep Erhart, who built one of the most successful and earliest seaweed businesses in Maine, bought their 1900s farmhouse with fifty acres of land down to Hog Bay in the town of Franklin in 1971. They'd met in New York City and fallen in love. They still talk about those years with that starry-eyed energy of the young. Prince Edward Island, in the Gulf of Saint Lawrence, was their dream place for a home. They had planned to move to Canada, not Maine.

"It was so beautiful," Linnette says. "We met people we liked. We found a farm we wanted to buy, and showed up at the border wanting to emigrate, but we couldn't get in unless we had jobs."

Shep adds, "Or ten thousand dollars. That was the deal. We had ten thousand dollars, but we wanted to put that into a house. The Canadian government was afraid we were going to go on the dole once we crossed the border."

"So we turned around and came to Maine," Linnette says.

They paid visits to Helen and Scott Nearing, leaders in the homesteading movement, and to Marjorie Spock, an anthroposophist and biodynamic farmer and the sister of Dr. Benjamin Spock, all of whom lived in Hancock County near the water.

"They were a big draw," Linnette says. "Marjorie was magnetic." Linnette and Shep had become followers of a macrobiotic diet while living in New York and wanted to grow and harvest as much of their own food as they could. "Originally we were buying the seaweed we ate from Japan," Linnette says, "but we were trying to live on three thousand dollars a year, and we didn't have the money to buy six-dollar packages of seaweed."

They read Euell Gibbons's books on eating wild plants and took account of his statement that most seaweeds in this latitude are not toxic. In a short time they found a place on the coast where *Alaria* grew. The seaweed looked familiar to them.

"I said, 'My God, that looks just like wakame!'" Linnette told me. It was an Atlantic species similar to wakame, one of the primary edible seaweeds of Japan. They picked it, took it home, and made soup with it, and that was the beginning: they became independent seaweed harvesters.

I am listening to this story, sitting in their new house, which they built after a fire burned their beloved farmhouse to the ground in 2007. They have designed a place that fits each of their particular needs, bringing to it what they love: plants, granite stones left over from an old quarry, a greenhouse space. They sited it at the back of their land, close to the water.

The evening light throws a soft chiaroscuro across the big field around their house. The tide is out in the bay, it's at dead low, and from the windows we can see the sheen of tidal mud as the sun goes down. A last ruby-throated hummingbird, a male, sips at their sugar-water feeder.

Shep is compact, with close-cropped hair and a gentle smile. Linnette's voice is thoughtful, articulate, clear as a bell. She grew up near her grandparents' farm outside Jacksonville, Vermont, where her mother, a single parent, worked as the town nurse. Over the years she has become a master gardener and a fine vegetarian cook.

Shep's great-grandfather, Charles Erhart, a confectioner, cofounded the Pfizer pharmaceutical company along with his cousin, Charles Pfizer, a chemist, in Brooklyn, New York, in 1849. Both were hardworking, smart businessmen who built this historic company from scratch. Shep had plans to become a doctor after he graduated from Yale, but in the heady counterculture world of the late sixties and early seventies he went to New York, met Linnette, and instead moved deep into the country. In 1974, Seraphina, their daughter, was born.

"We started to look around to see what else was edible here," Linnette says. "Then our macrobiotic friends began to call up and asked for a pound of this or that seaweed. We also put an ad in the *East West Journal* out of Boston, a two-dollar ad in the personal section, saying, 'Send us a couple of bucks and we'll send you some seaweed.' We'd get an order, go out and gather it, dry it outside or in the kitchen. And when it was dry, we'd package it up and send it off."

"In the beginning, one of the places we harvested was Schoodic Point," Shep says. "It's illegal to harvest there now, because it's part of the national park system, but back then it was wonderful. It sticks way out into the ocean and that's where *Alaria* grows best, in those turbulent waters. We'd go at low tide, sometimes at four in the morning or in the evening, when the moon is full or new, so the tides are bigger. Sometimes we'd watch the moon rise as the sun went down, sometimes the moon went down as the sun rose."

Linnette says, "We were getting beauty burns out there."

"Beauty burns?" I ask. "What are those?"

"It was gorgeous. Beautiful," she says.

"And we were lugging very heavy bushel baskets," Shep adds. "They were the old ash baskets, and we put a back strap on them. We'd fill them and put them on and stagger up over those big rocks to our Volkswagen. We trashed so many cars with the salt from that wet seaweed!"

"For every hundred pounds of seaweed," Linnette explains, "you get ten pounds of dry. When we dried it inside the old farmhouse, that ninety percent of water went into air and the wallpaper fell off the walls."

"Just peeled right off!" Shep says. "Drying is the hardest part." He pauses, remembering, perhaps, the look of the damp and drooping antique wallpaper they had inherited when they bought the place.

Linnette continues, "Later on, we'd look for a window of nice weather and we'd go down east and we'd camp on a beach."

"We'd put up clotheslines," Shep says. "We'd dry the seaweed on the clotheslines. We just stayed there and hung the seaweed up to dry and, while it was drying, go to sleep, and get up and do it again. It was a sweet time in our lives."

"But it was in the spring," Linnette says, "and it was a push for me because there was a garden to plant and a kid to take care of. I think until Seraphina was four or five, I felt in order to pull my own weight I had to lug as much as Shep did. I had to do as much work as he did."

Shep laughs: "I felt the same way. I had to keep up with her. And then there was the danger. Especially at Schoodic. The big southeasterly swells would come in from way offshore. We used knives or scythes and harvested in between the waves. And when we discovered that there's kelp right here close by in our Reversing Falls area, we'd go harvest it. We'd go out in winter."

"I always considered that more dangerous because you're standing in the hard current," Linnette says.

"Yes," says Shep, "dangerous and very, very cold."

They were harvesting *Alaria*, sugar kelp, dulse, and a species of laver. But they weren't making much money. They also made Christmas wreaths and picked blueberries, and Shep hired himself out for carpentry and odd jobs.

Then, in 1976, the University of Maine at Orono held an international vegetarian congress, and they rented themselves a booth. Linnette drew up labels, had them printed on white paper, and attached them to paper bags of seaweed. Shep painted a sign on a big piece of driftwood: MAINE COAST SEA VEGETABLES.

"We sold out almost immediately," says Linnette.

"That was big," says Shep. "Everyone was interested because vegetarians, when they are weaning themselves off of meat, crave minerals, and seaweed is a source of those minerals."

Linnette remembers that they built slowly, step by careful step.

"First," she says, laughing, "we did it in the house, then in the barn, then we bought the old salmon-processing plant and did it there."

"We've been in this business for over forty years," Shep says. "Sticking to something isn't necessarily the right formula for everyone, but it has worked for me—just staying with it—even though many times I wanted to get rid of the whole shooting match. We were also growing grain, trying to grow all our food from soup to nuts. One year we grew a whole field of sunflowers. We were going to squeeze the seeds for oil. We never got that going. We were doing some crazy stuff. We were strong. We worked hard. And we never took time off. It was years before we realized it was OK to take a break."

But in the eighties, Linnette told Shep she didn't want to do the seaweed business anymore.

"We were having too many fights about what the vision was. I wanted to keep it small, with paper bags and no advertising, no selling to distributors. He wanted to go the big route. I don't think there's a right or a wrong way, but I thought, This is going to be a business or a marriage, not both, and I think I'll hang on to the marriage part."

Their company and Larch Hanson's endeavor in the Washington County town of Steuben were the first start-ups in Maine that harvested seaweeds for food. Larch believes in keeping it small and in teaching the discipline of living within the limits of what a local harvest provides from the wild. His ideas were more in line with Linnette's.

For Shep, there was a commitment to the macrobiotic community. He didn't want to turn his back on them, and the number of macrobiotics who depended on Maine Coast Sea Vegetables was growing, which meant that the business had to grow with them. He had also received letters from nursing mothers, and people with cancer and other illnesses who were using the products medicinally.

"All this spoke to me," Shep says. One can't help thinking that his former commitment to doctoring had kicked in, in an unanticipated but powerful way, along with the press of a genetic thumb from the entrepreneurial genius of his great-grandfather.

The business, under Shep's management, grew. A leap forward came when Paul Hawken, owner of Erewhon, the first organic store in the country, contacted him.

"He wanted us to supply the store on a steady basis. That was a different ball game. We hadn't built an inventory. We just went out and harvested on demand," Shep says. "Now we started a whole other cycle—harvesting even though we didn't have a customer, because we knew we were going to sell to Erewhon at some point. I started hiring employees, and I needed more harvesters for a year-round steady supply.

"It took five to ten years to figure all this out," he says. "And seaweed is hygroscopic: you have to be careful because it soaks up moisture. If it picks up too much moisture, it can pick up a mold. So we had to learn how to protect it, how to store it. Then other places wanted it, and we got our first distributor, Bread and Circus. They had a truck that came up, and we'd put the seaweed on it, and they'd take it down and deliver it to health food stores along the East Coast."

"We sold to some of the earliest co-ops in the country," Linnette says. "We were selling to both coasts pretty soon, and hardly any in the middle, which is still kind of how it is today." She had stepped away from the business, but to her it was always theirs, and she kept her eye on the details.

"Part of eating seaweed is you've got this big long leaf. We used to sell the kelp as one big leaf rolled up in a bag, and when you opened the bag you had to unravel this thing. A lot of people said, 'What do I do with this?'" Shep says, laughing. "So that's when we started cutting them up into six-inch strips, and we'd tell them if you want to put them in your soup you snip them up, or if you want to put them in a salad you can marinate the strips and cut them up and put them in a salad."

By 1986, the business had gone from a kitchen-table operation to a $250,000 company, and it kept growing. Today it is valued at more than $3.2 million. As the business enlarged, step-by-step, they didn't take out loans but rather moved forward on what they made. The primary objective was to introduce the American public to something new; to make sure what they sold was clean, the best they could find; and to hire workers out of the local community who would stay with them because they were treated well. And they made money. Looking back, it seems inevitable and easy, but it wasn't.

A WEEK BEFORE, I had spent two days in the old one-story salmon-processing building that has been home to the business for twenty years. It sits along Route 182, just beyond the center of the crossroads town of Franklin, a few miles from their home. When I stepped into the vestibule, which was casually decorated with photographs and drawings of edible seaweed species and some of the products the company produces, I caught the distinct whiff of a low-tide beach—that sulfurous smell tanged with salt.

Under their ownership, the building had expanded into a warren-like series of dark hallways and little bright rooms where the processing and packaging and boxing up took place. But first I came to the large, open room at the building's center, where sacks of dried dulse were being laid out on white screening tables and picked over by a staff of five. I joined them, tying up my hair in a bandanna. The women all wore bandannas or hairnets, and we sat on stools, spreading the pungent, salt-laced purple dulse before us, snapping off an occasional holdfast, a few small periwinkles, tiny desiccated brine shrimp or mussels, an occasional infant green crab, and the ever-present epiphytes—smaller seaweed species that attach themselves to the growing dulse blades. We ripped away discolored portions and ragged edges and stuffed all this seaweed refuse into a bag that would end up in Linnette's compost pile.

The room was filled with the brittle sound of the dried seaweed being worked over in our carefully washed hands and the comfortable chatter of people around the table, a man and four women, who have held their jobs here for years. Afterward the dulse was packed into small plastic bags printed with the company name and other information for retail, and every package was weighed and pressed shut.

I stopped by Shep's tiny office, and on the way out I ran into his daughter, Seraphina, who is now the general manager of the business. She invited me to watch the cutting of sugar kelp, pointing me down a dark ramp, through a passageway, and into a room where a woman named Lauren was working, her hair in a hairnet. A small, friendly, capable woman, she drives down from Cherryfield, the river town about twenty miles northeast of here, arriving at 6 a.m. most mornings, and lets herself in. This is her room, off the moistening rooms, two large, dark closets where the kelps, which arrive dried from the harvesters, are laid out on screen shelves as a humming humidifier introduces enough dampness to make them pliant.

Lauren opened the door to one of the moistening rooms, where the long blades of sugar kelp lay on the shelves and a sweet-salt smell, a light and lovely perfume, filled the air.

She collected the softened kelps in her hands, cleaned them off with her fingers, and ran them through an old band saw so that each handful was six inches in length. Then she stacked the cut kelps in a cardboard box.

Leaving her, I visited Mickey, who works in another small space down another hallway. His room was crammed with bags of seaweed flakes and an oversize food-grade cement mixer. He was making a blended seasoning mix of *Ascophyllum* harvested here and *Alaria* granules sent down from Canada. (The word *Alaria* comes from the Latin word for wings: the wings are the sporophylls, the parts that hold the spores for the next generation. Linnette had told me that *Alaria* is one of the most delicious kelps, but the dried flakes in this room smelled a bit like wet dog.)

Mickey, completely at home in his job, scooped and measured precise portions of *Alaria* and *Ascophyllum* and turned on the mixer. A roar out of an urban building site erupted into that constricted space, and we both laughed as the machine whirled the stuff around. A tempest in a teapot. Eventually he turned the machine off. The room went quiet. He carefully scooped the blend into sacks to be packed into cylindrical shakers upstairs and sold to season a variety of foods—from soups to stir-fries to salads.

Mickey directed me to another small room, where Kara worked. She was stirring batter in a steel bowl with a large spoon. The mix is a combination of toasted sesame seeds, kelp flakes, brown rice syrup, and maple syrup, the ingredients for the company's popular Kelp Krunch bars. Linnette had told me it's impossible to stir Kelp Krunch batter using an electric mixer because it is just too stiff, and the edges have to be constantly scraped into the center. Kara is young and strong. She wore a stylish hat and moved with an easy grace around the room, which contained an industrial-size oven and cooking gear. She spread the mixture onto a large baking pan, rolled it out, and set it in the oven. As it cooked, we chatted, and she told me that she and her husband own a farm not far from here, where they grow most of their food and homeschool their kids. Soon the delicious smell of baking filled the room with a festive scent.

As I was leaving, Seraphina handed me a bowl of soup she had made with smoked dulse. She knows I am not a seaweed eater, but she wanted me to give it a try. Watching Seraphina watching me, I tasted the soup. It was a hot mix of soy milk and potatoes and corn and a big handful of smoked dulse, which Lauren had told me she adds to her weekly pot of baked beans.

It was good. "Good!" I told Seraphina, and she gave me an "I told you so" smile.

THAT NIGHT AT Shep and Linnette's house, Shep rises off the couch to put a log on the fire he's built in the fireplace. It isn't cold, but the fire is comforting, and as he coaxes it into flame, a neighbor calls to report two black bears crossing the Erharts' field. He sits back down and tells me that the dulse that grows on this part of the coast is tough and not especially prolific, but on Grand Manan, the Canadian fishing island that lies off the coast of far Downeast Maine, there's a lot of it, and the harvesters have taught themselves how to tenderize it.

"It's like the tea we get," Linnette explains. All black tea has been fermented, and on Grand Manan they do the same with dulse, packing it in boxes and compressing it. It is damp and salty, and as it begins to ferment, like sauerkraut, it gets especially tender, because it has started—just started—to break down.

"Everyone loved that dulse. They dry it really dry and then add a little moisture to it and let it sit for a few weeks, and it starts the process of tenderizing. It's only sun dried to begin with so the enzymes are still active, and those enzymes start digesting some of the proteins and the tough cell walls," Shep says.

"We began a relationship with Leroy Flagg, the King of Dulse, they call him up there. We buy his dulse. Pretty soon I was bringing back boxes of it from Grand Manan. Each box weighed about sixty pounds. Now we bring it in a container, but I can't get enough. I've gone over to Nova Scotia, which is another shore that has a lot of dulse on it. The

Digby area. I have a bunch of harvesters over there. I still can't get enough. We have to ration. We get about thirty thousand dry pounds a year, and we could sell sixty.

"A while ago I got involved in a scheme to grow nori in an aquaculture site in Eastport. *Nori* is the Japanese word for what we call laver. On this coast, we have four species of it, two of which are used for food, but they are not the Japanese or Chinese variety.

"The idea was backed by a Merrill Lynch guy who had a lot of money and thought it might be fun to have a seaweed company. He put about a million bucks into it. He hired two phycologists, but they didn't have a lot of hands-on experience, so they came to me to be the marketing end of whatever product they could produce. The goal was to make nori sheets similar to the Japanese and Chinese nori sheets. We tried for three or four years to grow really good Japanese nori up in the cold waters of Eastport. But we didn't try to grow the local species. That was our mistake. Ours is *Porphyra umbilicalis*—a tougher, thicker nori.

"Before we gave up, I went out to the West Coast and bought a nori-making machine that was sitting in a shed, unused. I brought it back in a U-Haul.

"The process is this: You harvest nori off the aquaculture nets, chop it in a big mixer with a lot of water so that it's a thick slurry. It's just like making paper. Then the machine

of manufacturers here in this country, people who made products out of seaweed or wanted to add seaweed to the products they made, and what they wanted was something that could be assimilated into whatever they were making. They wanted it cheap because they had to buy a bunch of other ingredients. They were doing value added, and they had to make some money on it at the other end. In Iceland, companies are able to produce a really high quality kelp product at low cost because they dry it using geothermal energy—and that's free. So we started selling a few pounds of that and then I went over—I don't like selling stuff if I don't see where it comes from—I went over to spend some time there, checking out how the company, Thorvin, harvests it.

"It's a big operation compared to ours. The company has an agreement with the Icelandic government. They have lease sites and they are allowed to harvest kelps in them. They have a boat and a grid system. They say, today we're going to harvest ten thousand wet pounds out of this part of the grid in a huge fjord, probably twenty-five miles long. They harvest in big quantities. They use a *scoubidou*. It's on a crane, on a cable, it goes down and twirls and snags the kelp. It's pretty crude and it does a pretty tough number on the kelp. They get a lot of holdfasts, so they only do it every five years or so. They leave buffer zones, so let's say they harvest a patch the size of this room, then they'll leave patches all around it and the idea is the spores will come into that cut area and resettle it.

"They've been doing this in the same fjord for about thirty years now."

IN 2011, TWO events changed the basic scope of the seaweed business. The first was the Fukushima nuclear accident, in March, which caused the US West Coast's appetite for safe, edible Atlantic seaweed to skyrocket, because the people who might have bought from Japan or California were afraid of contamination spreading across the Pacific. The second came five months later, in August, when Hurricane Irene scoured the western North Atlantic and washed wild seaweeds out to sea and the year's harvest of many species vanished. They just disappeared.

What Shep realized from this double punch was that the Atlantic seaweed industry was bound to grow, and that for a steady, reliable supply, they'd have to start to move away from wild stocks to farmed seaweeds to meet the increasing demand. In the past few years, he has begun a partnership with the University of Maine to start an aquaculture business in Frenchman Bay, a short drive from Franklin.

"We may be at the threshold of a sustainable harvest of the wild crop," he says. "It makes sense: if we're going to meet increasing demand, we start to cultivate it."

This winter and spring they plan to set out kelp, *Alaria*, dulse, and nori to begin to learn how to grow them in these nearby waters. And to accommodate the growth of the business, Maine Coast Sea Vegetables will take leave of the old salmon-processing building and move to a brand-new $1.5 million facility in the town of Hancock, twenty miles away.

A MONTH AFTER the move, I toured the new facility with Seraphina, who showed me every nook and cranny of its 17,600 square feet.

"Brand new doesn't mean perfect," she told me as we walked. "We've had some glitches, nothing unworkable, but we have a way to go to be the best we can be here."

Now Lauren cuts kelps in a room with a window that throws light all day. Her drive is a bit longer, but it's fine, she says, and she enjoys her new space. So does Mickey, who makes the seasonings in a larger, better-equipped space with two cement mixers. And Kara has a brand-new kitchen with a big sink and a new induction burner to heat up the rice syrup for the Kelp Krunch bars. The moistening rooms are larger and can be converted into dehydration rooms for seaweed that occasionally arrives damp and needs to be dried before it can be processed.

Shep has his own room for a hammer mill, which Seraphina laughingly calls his "new toy": a huge grinder and sifter capable of processing seaweeds into flakes and flour. It stretches across the entire room, hung from the ceiling, with a silo and four white collection bags.

"After Fukushima, some people were looking to the Atlantic for clean seaweed. Once the news died down, we had less pressure on us, but there are still people who won't buy from the Pacific, even the US side of the Pacific. We have different species of seaweeds here than they do there. They may be in the same family, but they are not the same. For instance, the seaweed *Alaria* we've been calling wild Atlantic wakame. But do we still need to do that? Wakame is an edible Pacific seaweed. It's closely related to *Alaria*, but it isn't *Alaria*. Perhaps it's time for us to let Atlantic seaweeds stand on their own."

We sit down in the new conference room, and I ask Seraphina about her experience as a child of parents who went "back to the land." She smiles because she knows that my children grew up in this same movement, which brought us out into the country, with all the challenges regarding how to live.

"I grew up going to the shore and playing in tide pools, and I recognize now what a privileged childhood that was. I had lots of time outside." Seraphina is forty-two now, energetic, welcoming. As the company celebrates its birthday this year, it's been forty-five years since the day her parents first processed seaweeds to sell in their kitchen.

"Linnette tells me that when I was four, people would come to the house, and we were harvesting, and I wanted to give them the seaweed tour. There was a pride in it for me.

"But when I turned ten, as an only child, I asked to go to public school. I had a weird name, and I came to school with wacky vegetarian lunches, and kids can be cruel. It was hard. I ended up graduating from a private high school in New Hampshire, which was a little more open, but when I went to college, I wouldn't tell anybody what my family did. So everybody guessed crazy things. By the time I finally owned it, most people said, 'Oh, that's cool!' That was a first for me, and I started to embrace it. I started to see good things about how my macrobiotic parents raised me.

"When I began here at the business, I worked from the ground up—the equivalent of working in the mailroom. I started in production because I wanted to earn the respect of the people who work here. I didn't want anyone to think that this is being given to me, that I hadn't earned it. It was also important for me to get to know and to learn everything about the business and the products and what we build our finances on. I had already gone off on my own and run a photography business. That was helpful, but I needed to learn about running this particular business.

"One of my biggest challenges when I started was that I didn't actually produce anything. Like a beautiful photographic print. It was hard for me to remember, sometimes, the value of my day, because I didn't have something clearly to show for it. My first couple of years, the struggle was in figuring out that value.

"But I love the collaborative vision we have. I go with Shep every year to talk to our harvesters. I'm building a relationship with them. Shep's the visionary and I'm more of a people person. There's a generation shift that's starting to happen, and I don't know what the future is—but I want the business to thrive."

WHEN I LEAVE Seraphina and walk out across the broad new parking lot to my car, I can hear Shep's voice, from the night we sat in his home before the fire and talked about our lives: "We've borrowed money for the first time," he said. "It's a little bit scary. If we get an oil spill out in the Gulf of Maine, we're toast."

# CHAPTER 8

# Seeding the String

Sarah Redmond is a farmer, and this morning, early, with fog hanging in the branches of the softwood trees and drifting across the road, I am driving down east to the farm she tends. It's off the town of Sorrento in Frenchman Bay, with a view across the water to the mountains of Acadia National Park. When I arrive, the bold, rounded shapes of the mountains are hidden behind the fog bank, and long horizontal strands of fog drape the inner islands.

Sarah has rowed her dory out on the bay and is unmooring the motorboat that will take us to the thirty-acre patch of clean water just where the bay drops to eighty feet deep—that is her field. Within it, demarcated by buoys and blaze-orange moorings, she grows seaweeds for food. This entire bay, with its ample currents and tides, is an enormous bowl 150 feet deep at its deepest point, which means seaweeds, both wild and farmed, are constantly bathed in cold, nutrient-rich water.

Last fall I met Sarah for the first time at the Center for Cooperative Aquaculture Research, where she works for the Maine Sea Grant program under the auspices of the University of Maine. She keeps an overflowing office, full of seaweed books and projects, in an old trailer off to the side of the compound. The center was founded by the University of Maine in 1999 on the shore of Taunton Bay. It houses almost everything you can think of to raise and study fish, invertebrates, and sea vegetables as aquaculture projects, a resource for scientists and students and for fishermen hoping to create new jobs along the coast.

I got there before Sarah did and sat down on the wooden steps of the trailer to wait. When she drove up in an old white truck, she made a tight spin that raised some dust, parked, stepped out, and slammed the door shut behind her. I was surprised to see that she looked as if she had stepped straight out of a Jane Austen novel: pale, unblemished skin; long, straight black hair and dark eyes; a lithe, almost delicate Austen shape. As a counterpoint, she wore heavy boots, jeans, and a hoodie and was prepped to talk seaweed. She's a leader in the state's energetic venture to cultivate and grow native seaweeds out in the bays where they still grow wild.

We went into her office and sat down.

"Everybody's talking about food security, but nobody's talking about nutrient security," she began. "People aren't getting the nutrients they need, and seaweed can do that for them in small quantities, by folding it into food that people already eat."

Sarah was born thirty-six years ago in Litchfield, a small inland town near Augusta, the state capital.

"When I was growing up we had a large garden, and as soon as I was old enough, I was in charge of it. Every year. I spent a lot of time out there and it made me happy."

In high school, someone told her she had to figure out what to do with the rest of her life. It seems a bit early for such a decision, but she took it seriously and put together her love of gardening and her love of the ocean: "I said I want to be a seaweed farmer," she told me. "Then I went to college for aquaculture, and when I graduated, no one was growing seaweed in Maine."

She worked for a few years as a fisheries observer out on commercial fishing boats. But it wasn't what she'd hoped to do, so she took a job with Tollef Olson, a pioneering aquaculturist in southern Maine. From there, she received her master's from the University of Connecticut, studying kelp aquaculture under the internationally renowned seaweed expert Charles Yarish so that she could come home to Maine and initiate kelp farming here.

"I was hired here before I had finished my master's," she said, and she's worked with seaweeds and seaweed farming ever since. "It's as if seaweed just appeared in Maine and just showed itself to the world. Somehow we couldn't see it before, or maybe only some people could see it, but it's becoming mainstream, which is great. At the same time, we have to figure out how to cultivate it. If the whole world suddenly wants it, we need to put safeguards in place so that we're not wild-harvesting too much too fast.

"I want to develop culture techniques on the farm so we can offer an economic change for fishermen who are running out of options. When people know what we have here and why it's valuable, and that it's a natural resource, we'll create a system that will be good for people. All over the world, big companies are going into other countries and buying up their resources. There are these giant global corporations investing in food and water because they know that's the future, but they've caused complicated issues locally," Sarah says. "One of the most important things about this work is having local people reconnect with the natural world. Whenever it's reduced to theoretical arguments, or to an industry framework, we're not really talking about the seaweed. The best part for me has always been just being here."

Her father, a man with an inventive, creative mind, has built her a geodesic dome out in a patch of woods, where she lives year-round. It has solar panels to run lights, and no running water or refrigerator.

"It's the best place to sleep in the world," she said. "It's my cave."

TODAY I WAIT on the dock for her to bring in the flat-bottomed boat. I am standing alongside Dr. Susan Brawley, the internationally known phycologist from the University of Maine, two of her students, and Shep Erhart. The cool air smells of salt and iodine. A strand of wild sugar kelp floats by, its stipe, filled with oxygen, as buoyant as a life preserver. It trails a twelve-foot-long blade. I lean over and pull it up onto the dock. Susan Brawley tells me it's typical of the wild kelps that grow underwater here.

Around us, lobster boats rock slowly in the tide, tugging at their moorings. It's late May, and the sailing crowds have not yet launched their pleasure craft. This is still what we call the working waterfront.

Sarah brings her boat up to the dock, and we climb in. We head out, away from the mainland behind us, the tops of the spruces and pines pointing through the fog. This seaweed farm is an experiment, although the harvests, which are plentiful, are already being sold as food. It is leased by the state to a local fisherman who is working with Sarah and Shep to build seaweed aquaculture into a commercial venture. They are learning how to grow the best native edible seaweeds they can. The farm requires no alteration of the surrounding environment other than the moorings and the ropes and buoys, all of which are removed in the summer.

THE FOG IS leaving the mountains as we head to the site. It seems as if the mountains rise straight out of the water, filling the horizon, reminding me of scenes of Japan or Ireland: water, mountains, everything spare, green and blue, with leftover wisps of gray-tinged fog. To work in beauty is a gift that few people are able to enjoy, although one might make the argument that it's a human birthright of sorts, working where the sun rises, the day clears, the smells of the wild world are all around you, and you are a part of something very old, very big. When I mention this to her, Sarah laughs and reminds me that she's been out here all winter working in snow and serious cold.

Today she says, "Wow! It's pretty nice!"

This farm, along with three others southwest of here, are the first full-time endeavors in the state to grow and sell seaweeds for market from Maine bays. Sarah has also worked with a number of shellfish aquaculturists who want to include seaweeds on their farms as well, a process called integrated aquaculture, which is the way the future will probably go.

"We don't want to be monofarmers. We want to learn from our agricultural predecessors about multicropping," she had told me back at the trailer. The concept of growing shell-fish along with seaweeds, all of them relatively clean ventures without the contaminants caused by the feed, supplements, and fungicides used in most finfish and shrimp operations, has attracted a number of innovators here in this country and overseas. Borrowing the idea from farming multiple crops on land in a single field, these people are concentrating the natural processes of the ocean, in which mussels, clams, and oysters siphon in suspended particles from the water column and release dissolved nitrogen and phosphorous. Seaweeds then incorporate the nitrogen and phosphorous into their tissues to help them grow, much the same way nutrient cycling works in a kitchen garden, with lettuces thriving on a composted cow manure supplement. This overlapping farming is a clean way to use the oceans for food.

"I've been acting like this seed giver," Sarah told me. "I give people seaweed seed and they put it out on their shellfish farms and they can play around with it. And I've been getting a lot of interest in kelp farming from lobstermen because aquaculture kelp is a winter crop. They could have a kelp farm in the fall when they take out their lobster traps. Then they harvest the kelp in the spring before they put their traps back in."

Here, where the water, though warming, is still bone cold during the winter seaweed-growing aquaculture season, most of our bays are essentially pristine. Every bay in this state is different, however, with somewhat diverse bottoms, water temperature variations, differing amounts and types of particulates and dissolved minerals suspended in the water column, and different currents, tidal lifts, and other sea life, including wild seaweeds, all of which can potentially affect a crop.

You've got to choose your bay. And you've got to choose your seaweed species and match them to your bay. A seaweed farm would not be sited above a wild kelp bed, which requires hard substrate to attach to, structures like cobble and rock. Most of our coastal bottom is often mud, where wild seaweeds won't grow, but suspended farmed seaweed will. Growing seaweed over mud is good husbandry because it doesn't impinge on wild habitat to create a cultivated system. Farming seaweeds—as long as native species are the ones that are grown—should not have a negative impact on the environment in the bays, and sea farms create structures, which quickly become good temporary habitats that attract other marine life.

SARAH DRAWS THE boat up to the first seaweed rope, leans over, and hauls a loop of it above the water. It shines with a dense growth of sugar kelp. We all lean over, admiring the lush abundance, wet and sparkling in the morning light. And then we pluck a blade, rip it into shares, and eat it. It tastes sweet and salty, and it's crisp, easy to chew. We check the lines of *Alaria*, pull one up, and pluck a few sporophylls, the reproductive wings at the base of the blades, and eat them, too. This is turning into a bit of a feeding frenzy as we move on, tasting as we go. Sarah pulls up a rope of dulse, dark red and shining. As the seaweed rope rises out of deeper water, its color lights up the surface like a line of votive candles.

"We're trying out companion plantings of dulse on some lines with kelps," Sarah says, "because the dulse needs shade or it will bleach, and the kelps keep it shaded." She moves the boat ahead, and lifts a rope where a form of sugar kelp she calls skinny kelp, or *forma angustissima*, is growing.

Susan Brawley takes the rope and gives it a quick yank, then flips the kelp upside down so that the holdfasts are exposed. She shows us tiny movements within them. They are caprellids, small amphipods commonly called skeleton shrimp, moving their legs from within the rigid holdfasts. Slender bodied, they live in the holdfast filaments of seaweeds along with hydroids and bryzoans. This is the microbiome, made by the seaweeds themselves: small, complete places of interrelated lives.

In 1984, Brawley was studying seaweeds in China, and she devised an experiment in which she set some lines with caprellids and left some without. She found that the skeleton shrimp clean the seaweeds of epiphytes—seaweeds that grow on the host blades and cover them, rendering them unfit for a human food crop. Knowledge such as this, detailed and specific, the result of paying careful attention to lives that are quite small, expands our understanding of how things work underwater and teaches us that not all creatures who take refuge in farmed seaweeds diminish their value. Some, in fact, like these tiny creatures that look like they're waving at us, enhance it.

Shep asks her if these farmed seaweeds, which absorb carbon dioxide as they grow, serve a significant sequestration function in cleaning up the water and the air. Brawley says that in the larger sense, they are too small a portion in this bay, and this bay itself is too small a portion of the coast to make a measurable difference.

However, research scientists from Bigelow Laboratory for Ocean Sciences in East Boothbay, about a hundred miles south of here, have just completed a study on sugar kelps and carbon sequestration. What they found is that around a three-acre kelp farm, the seaweeds were taking in a surprising amount of carbon as they grew, creating what the scientists call a halo effect of good water quality as the acid levels dropped. Brawley is right, the effect can't alter the acidity of an entire bay, and it diminishes as the seaweeds slow their growth rate, and reverses as they begin to deteriorate. But by harvesting the sugar kelps at peak growth and then quickly reseeding the ropes, a small but steady amount of acidification remediation can continue throughout the growing season.

What this means is that many life forms, from calcareous algae to amphipods, isopods, shrimp, shellfish, and other species—anything that requires calcium to live—may find a refugium within this halo of water created by farmed kelps.

FOR SARAH, UNLOCKING some of the secrets in the nursery where she prepares the seed is as vital a part of her mission as growing them in a patch of open water.

Last fall, on a day of extreme tides, she had driven to the town of Harpswell to collect a particular morph of wild skinny kelp from a bay at dead low tide, because she wanted to try to seed it on strings in the nursery and set it out on the ropes here in Sorrento to see if it would grow true to form. And it did.

People like Sarah, searching for wild seaweeds that present valuable characteristics, travel to remote bays looking for variants, which they call forms, and remove samples to grow what they find in the laboratory. This sort of pioneering provides an inventory, a "seed bank." When she brought back some *Alaria* from a collecting trip, she invited me to help her get them ready to seed. In fact, they are not seeds but spores, which means they are evanescent and extremely fragile.

It was late September when we entered a building at the center the size of a small airplane hangar. She took us to a room with a metal table and brought in a stack of small pieces of dark brown *Alaria*, the size of spinach leaves with the feel of wet leather. We washed our hands, dipped them in a bleach-water solution, and began to prepare the blades.

Standing there at the stainless steel trays under fluorescent lights, she and I cleaned the *Alaria* blades with razors, scraping away layers of bacteria and microalgae we couldn't even see. We rinsed the blades, wiped them, then patted them down with iodine and set them on paper towels to dry. It was a long process.

In advance, Sarah had prepared immaculately clean seawater, running it through a series of finer and finer filters and submitting it to UV lights "because all of our fouling issues are microscopic," she said. She will release the spores of these *Alaria* patches into glass jars of the stringently clean seawater, where they settle on spools of string she has set for them within the jars. They look soft and fuzzy, like the nap on a baby's blanket.

"So you have billions of these tiny microscopic spores swimming around, and they want to settle. They settle on the string and immediately make a little germ tube and create their first cell, and start to become males and females," she explained. This is called seeding the string. The spores release gametes, which produce zygotes, which then produce sporophytes with tiny holdfasts.

The seeded strings will be set out in the farm in the bay to grow during the winter season, and the harvesting will be over by early June, before "everything starts to wake up and settle all over your crop—diatoms, epiphytes, everything," Sarah said. "It's a mess."

"When you seed the line, you wrap the string very tight to the ropes," she said. "And they'll step off the strings and attach themselves to the ropes and start to grow as seaweed."

TODAY, OUT ON the bay, we come upon a bountiful *Alaria* rope, the result, Sarah tells me, of our September work.

We move in the boat around the farm, and Susan Brawley and Shep and Sarah discuss this year's harvest, including how to deal with some of the problems that arise, such as the loss of the nori crop the first time Sarah tried to grow it. Nori, or laver, is usually seeded to nets because it is thin and fragile and they give it more structure. Sarah's nets became fouled with other organisms, and she says she might prefer to grow it instead in a tank back at the center.

Susan Brawley tries to talk her out of it. Nori, she argues, requires a great deal of care and handling—raising the nets, cleaning them off, setting them back at the correct depths— all of which means that Sarah and her crew would be spending a lot more time here in deep winter. But the nori would be a healthier, better-tasting crop grown out here. Sarah suggests trying to seed it to vertical lines tied to the ropes instead of nets. They'd be easier to manage. Brawley agrees it's worth a try.

I have sat around tables in the homes of friends who are farmers, and listened to the same kind of back-and-forth about their crops: what works, what doesn't, what to do about the next season, the risks to take on with new projects, the endless challenges. That's what farming's about. But this conversation, which is now turning to the weight of the ropes and the density of the growing seaweeds, and whether, in a few days' time, to cut them off the ropes and stack them on board, or to bring in the ropes with the seaweed still attached, is something more: it has a cultural and political edge.

These people in this little boat are, I believe, a vanguard come back to reclaim a big place—this Gulf of Maine—and to repopulate it with small farms, an old model, a land-based model, and a brand-new way to heal our relationship with the ocean. As they talk, a morning gust ruffles the surface of the water around the boat, and the boat rocks gently in response.

After a time, we continue from one mooring to the next, pulling up the ropes and checking them as the conversation continues about how best to handle the harvest, which is shockingly bountiful, and how to dry all of it quickly while it's still fresh.

"Lobstermen are used to catching and selling at the dock," Sarah tells me. "We can't do that. What we need is some infrastructure to dry big amounts of it—to turn it into something we can use. We're working with engineers at the University of Maine to see if we can set up a greenhouse system where we can do that."

But for now, she'll hang it up to dry on lines—as one would a week's laundry.

WHEN WE STOOD at the stainless steel trays this past September, Sarah talked about the implications of growing seaweeds. "It's important for people to think about the systems they want to build for the future and to be a part of, and it's important to realize that we have a choice. The big industrialized food system is not the way we want to go here in Maine. What I want to do is to create a new food system, a new model that works for the people who live here."

One might chalk these words up to a passionate enthusiasm that can easily be crushed by a sophisticated and powerful food industry. Maine is fast becoming the largest resource for edible and commercial seaweeds in the country, which also makes it vulnerable to takeover from the outside and to harvesting rampages from within its own ranks. But every individual who plants a seeded string that Sarah prepares and donates, and who is engaged in trying to make a living in a manner that, if care is taken, does no harm to the ocean, keeps that relationship between a person and a place on the water alive.

Some years ago, driving on Interstate 80 through southeast Nebraska in early spring, I remember watching the combines as big as buildings moving across the bare fields, a billow of brown dust rising behind them, and the darkly barren, unplanted fields stretching forever across that flat land. Every bit of land was appropriated for these industrial farms, these nations of soybean and corn. Only where the thin slivers of streams still ran was there any uncultivated brush at all. No trees, just a scrim of low bushes, and sometimes a red-tailed hawk circling above one of them, looking for something to eat. But to get off the highway, and onto the side roads, was to discover small, independently owned farms, not many of them, but a few, bursting with birdsong, lush green places where families still owned the land and eked out a living. I remember the fence posts and the meadowlarks perched on them, singing their hearts out.

When I look at the photographs of inshore industrial sea-weed farms along the Pacific Rim, especially along the China coast, I see Nebraska: huge, shallow bays extending for miles, filled with ropes and nets, all neatly placed in meticulous rows, row after row, and between them just enough space for the motorboats that the workers use to oversee the crops, some of which contain species introduced from elsewhere. The water is turbid and still. And in some of the farms, the bottoms have been scraped clean of endemic life.

Beneath farms such as these, which Toleff Olson tells me can be seen from space—they are that big—sediment begins to build up and nutrients begin to disappear.

Toleff Olson was one of the first people to bring sea-weed aquaculture to Maine, and he's a tireless spokesman for independent aquaculturists growing native species in ecologically sound projects. When we sit down to talk about seaweed farms, in Maine and beyond, he begins, "We've got the advantage here of seeing the mistakes that have been made other places, and we're learning from them. And we've got a strict set of guidelines. Our aquaculture laws are pretty well defined, so we'd never have, for instance, overcrowding, and we'd never be allowed to interfere with a sea grass bed or anything like that.

"You're under a microscope when you go for an aquacul-ture lease. It's hard on the business plan, but it guarantees the protection of the resources and keeping this coastline open and usable for everybody. The burden of proof is on us. Introduced seaweeds—absolutely not. There are very strict laws about that and there are not going to be any foreign species grown here. When you go for your lease, you have to show the source of your seed, and there's no way they're going to approve growing a possible invasive."

He tells me that in Maine, aquaculture farms value the spaces between crops because they protect and promote the natural water flow of the bays. Seaweeds, he says, need water flow to feed, and without it they begin to starve. If you want to grow a big, robust crop, you don't want to cram it all together.

"There have been some horrible practices, especially in some places in the Pacific, although I don't want to single them out," he tells me. "The mangrove cutting started with the shrimp farming, but they do it sometimes for the seaweed farms, too. In the Philippines, where they're taking shallow bays and covering them with nets, they can suffocate and kill the corals, or they might remove them to keep their nets in shape, and in some farms they've overplanted to the point where they now add fertilizers.

"In Europe they are doing seaweed aquaculture bigtime, from Norway down to Portugal. They appear to be using proper husbandry. I haven't been to all the places, but what I've seen is very good.

"It's easy to spread your crops out here," he says, getting back to the coast of Maine. "We have three thousand miles of coastline, not counting the islands, all these beautiful watersheds. As demand for seaweed increases, aquaculture production will increase, and areas that are underutilized now will be used, making the ocean more productive and adding economic benefits up and down the coast."

Recently, the US Department of Energy has awarded a $1,321,039 research grant to the University of New England, located in southern Maine, "to develop the tools to enable the United States to become a leading producer of macro-algae," with a focus on developing transportation fuels.

When I spoke to Adam St. Gelais, an assistant research scientist at UNE who is the project manager, he told me that grants such as this one have been handed out to institutions along the West Coast and up into Alaska, down the Atlantic coast south of here, and along the Gulf of Mexico. These projects are focused on figuring out how to grow native kelps in federal waters, three miles offshore, in winter, in open ocean conditions. The seaweeds would be hung from huge platforms, creating vast aquaculture farms, Adam tells me, although now such high-tech farming "lives mostly in researchers' heads and in computer models."

What does this portend for whale migration, for instance, and for the seasonal movements of schooling fish? And will the industrialization of the offshore affect the work—and the dreams—of the small, independent inshore aquaculturists like Sarah? We don't know.

SOMETHING SARAH REDMOND said stays with me as I think of what the future might hold: "Our system among wild harvesters is to go out and take. There's no seeding, no giving back," she said. "But here's an idea, an opportunity: seaweed harvesters, those who cut the wild seaweeds, start to reseed the beds in the ocean and get new populations going. Why not? They take. They give back. And if it works, they harvest again.

"Up in Canada, they took reproductive dulse and laid it out on rocks and boulders, and put nets over it, and allowed the spores to just drop onto the rocks. And it worked.

"There is so much potential here in Maine. We could be the new Japan. There's the increasing demand for sea vegetables; there's more awareness and creativity with our chefs now, so you're not just eating seaweed in a sushi roll. I'm thinking that Maine can be known for its blueberries and lobsters *and* its kelp. And that's where I want to be."

# CHAPTER 9

# The Uneasy Art
# of Making Policy

*Working with What We Know*

Dr. Brian Beal looks out his kitchen window onto Larrabee Cove in Maine's Washington County. It is early morning, low tide, and he is watching half a dozen clam diggers with hoes and hods, bent over and working the mud and cobble beyond the shore. Each has his or her own rhythm, thrusting the hoe into the mud, cutting a fan shape, turning the mud over and opening it up with quick chops, picking out the mature clams, tossing them into the hod, and stepping forward to drive the hoe in again.

Brian is tall, red haired, a man of unflagging energy, both physical and intellectual. He and his wife, Ruth, live in the house once owned by his grandparents, where he spent time as a child learning many things from his grandfather, such as how to dig clams, set a lobster trap, manage a boat, make a wooden buoy, and tie knots. He especially remembers how to pitch a clamshell into the air over the water, where it glides like a Frisbee, comes gently down, and rests on the surface.

If lobster fishing had been as lucrative a profession when he graduated from high school in 1975 as it is today, Brian says, he might have gone into it, as did most of his classmates. Very soon they were getting married, mortgaging boats and homes, having children. Instead he was living with his parents and attending the University of Maine at Machias. From there he went to the University of North Carolina at Chapel Hill, and then he earned his PhD in marine bioresources from the University of Maine at Orono.

Beal is an old family name in Downeast Maine, where records confirm that in the year of 1774, Manwarren Beal Jr. settled Beals Island. The county hugs the border with Canada. It may be among the most beautiful and wild places anywhere, but it is also one of the poorest. People have been scraping a living from the forests, the blueberry barrens, and the ocean for a long time, and there's nothing easy about choosing a life here. Brian was brought up on the mainland side of the Beals Island bridge, in the town of Jonesport, and the depth of family history informs his pursuit of science.

"This is home," he says. "What I ask myself is, how does the economy spin right here? How does it move? Well, it moves because of the resources out in the ocean.

clam hatchery, and in 2000 he won a Fulbright to study at the National University of Ireland Galway.

In April of 2015 he was honored with the first Bourne-Chew Award from the National Shellfisheries Association for outstanding contributions to education and for "promoting understanding and knowledge among industry and the academic community."

Brian's research is directed specifically toward outcomes that enhance the value of wild, harvestable inshore species. "The work allows me to scratch those intellectual yearnings I have, but at the same time what I learn has got some application for somebody's livelihood. That's why it's called *applied* marine research.

"When I hear the word 'experiment' it's like lighting a fire under my ass. It gets me going. I love to find out about things that nobody else knows. I love to design the experiment to answer the question. I'm good at it. And when we get the answer it's usually not wishy-washy, because we've done a lot of work to make sure we're confident in the results."

In 2010, Acadian Seaplants Limited, the Canadian seaweed company that cuts *Ascophyllum* in Washington County, came to Brian in his role as research director of the institute to ask for help. Their marine biologist wanted to conduct experiments to test the impact of harvesting

*Ascophyllum nodosum* on invertebrates and other seaweed-dependent wildlife along the Washington County coast. Acadian Seaplants is the giant international company with a large research department of its own. It's been harvesting *Ascophyllum* in Maine since 2004.

"I'd had serious doubts about any harvesting of *Ascophyllum*," Brian tells me. "But they paid for a student to help me, and I designed an experiment. My agreement with them is that I get to write up the results and publish them. We took the harvest all the way down to thirty-seven percent without seeing any impacts on invertebrates in the long term. We didn't sample mobile species, such as fish. We didn't sample ducks. We measured periwinkles, isopods, mussels, and green crabs—fauna that's in the seaweed at low tide. Amphipods have become a major story line as being part of the prey base for many of the ducks, and we didn't see the drop with them that was supposed to happen."

The amphipods Brian speaks of are those laterally compressed shrimp-like creatures, many of which inhabit the briny seaweed windrows tossed onto the upper shore by storms, while others live in the seaweed tangles in shallow water. They are scavengers of everything tiny and disintegrating and consume both dead and living seaweed tissues. Some are stationary, or sessile; others scoot sideways through the algal beds with surprising speed. They are a vital part of this community: the detritus of the near shore is their food, and in turn they are eaten by many.

IN 2013, THE Seaweed Council, the consortium of seaweed harvesters and processors in the state, approached the Maine Department of Marine Resources requesting that their industry be regulated. The department's stated mission is "to conserve and develop marine and estuarine resources; to conduct and sponsor scientific research; to promote and develop the Maine coastal fishing industries."

In reality, there is little money to set aside for research or for studies of conservation techniques, and the department through the years has overseen the bust of more than a handful of valuable fisheries. What is written up as a mission is frequently overridden out in the field by rapacious fishing techniques, the inability of ocean scientists to assess a plunge in fish stocks quickly before it becomes a crisis, fishermen competing against one another for diminishing resources, and pressure on the world's coastal communities to supply fish—and lots of them—to people everywhere.

When the seaweed harvesters and processors came to the department for help, the department was handed an opportunity to work with them as allies to prevent seaweed overharvesting and to avoid the potentially dispiriting and vexed job of trying to build back a damaged resource in the future. In March of that year, the department invited Brian to be part of the group to set policy on a single seaweed species, the one that bears the brunt of harvest: *Ascophyllum nodosum*.

"When the seaweed harvesters approached the Department of Marine Resources, they wanted a management plan," Brian explains. "Something like that's never happened before. It's the first fisheries management plan that's ever been initiated by the people themselves working together with the department. It's always been a top-down approach here: 'This is how we're going to do it.' But this time it's different. This is 'Let's get together and figure this out.' I liked that."

The consortium's request carried a degree of self-interest, of course, because companies and individuals who are making a living from *Ascophyllum* in this state now don't want to be blindsided by runaway cutting pressures from outside their ranks. One way to control that possibility is to make rules.

"They did it because they knew there were going to be increases in harvest, increases in the rate of cutting and in the number of harvesters," Brian says. "That, I think, was the impetus."

In October, I drove inland along the Kennebec River to Hallowell, a river town downstream from the capital of Augusta, to the Maine Department of Marine Resources. I had been coming regularly as a member of the public to sit in on the meetings to develop a statewide seaweed policy. The working group, seated around a long rectangular table, was composed of three scientists, two conservationists, two seaweed processors, one harvester, and three members of the department. These people were responsible for sifting through a daunting amount of information—scientific and otherwise—to arrive at a feasible, implementable plan for harvesting just one species of seaweed. Behind them sat members of the public.

At that point, the group had only a few more months to come up with something to offer the department, which would then take it to the legislature for approval. I thought the people around the table were looking a touch paler now, and the exchanges in the room occasionally turned terse. The group was receiving a fair amount of pushback from the public, some of whom were attending every meeting, either because they didn't want more restrictions on their freedom to cut *Ascophyllum* (the law already stipulated that they had to hold a license and comply with a few minimum regulations), or because they didn't want any seaweed harvesting at all because they were fearful that we were, indeed, fishing down the food chain and had reached the bottom, or because they were worried that the outcome was going to favor the harvesters. Maine is a big state, and many in the audience, and those in the working group, had driven a long way through the dark of early morning. A number of people in the metal chairs along the wall had already spoken as the day wore on, and it was lunchtime.

The subject at this point was how deep a cut could be made to *Ascophyllum*, measured from its holdfast to its tip, without compromising either its ability to regrow or its contribution to the immediate habitat. This may sound esoteric, but it's essential both for ecosystem health and because what the people at the table eventually decided would affect how harvesters do their work, as well as the work of wildlife biologists and marine patrol officers. Cutting the blades of *Ascophyllum* that stand upright in a full tide clearly has an immediate impact, but how much, and for how long, is a matter of disagreement.

Later that month, Brian and the research director for Acadian Seaplants took the working group out in boats to see one of their experimental *Ascophyllum* plots, one that they had cut only a few months before. "When we asked them to tell us which was the control and which were the sites that had been cut," Brian says, "they couldn't do it."

By January of 2014, the group had a completed plan, incorporating six management recommendations. The first was that *Ascophyllum* could be cut by machine, rake, or knife, taking all but the lower sixteen inches of the blades. They agreed that this length should preserve the regenerating capacity of individual seaweeds and protect the general canopy of the stands.

The second recommendation was to give sectors of the shore to individual harvesters or to companies for up to six years and renew the licenses only if they kept to statute and rule. It did not restrict sector size. Depending on the needs of a harvester or a company, they might choose large or small sectors, but only 17 percent of the *Ascophyllum* within the sector could be cut yearly.

The third recommendation was that another group be convened to designate no-harvest areas to protect sensitive wildlife, especially bird habitat. The fourth recommendation was that this plan be implemented for the entire coast. The fifth, that new harvesters would be required to take a training workshop. And, finally, in five years the plan would have to be reviewed.

At the end of the written document of recommendations, there was a page stating that more research needed to be done on *Ascophyllum*. We haven't learned enough about what part it plays in the inshore life of many species, and how cutting may or may not alter subtle but important aspects of that relationship. Although a list of scientific studies was enclosed, there remains a surprising element of uncertainty.

WE ARE SITTING at Brian's kitchen table. It's past noon. The sun is withdrawing from the big windows to the east, and outside a few flakes of snow have begun to slide across the air. On the wall behind us hangs a hand-drawn map of a tiny island, just a clamshell's throw across Moosabec Reach from Beals Island. Sketched long ago by Brian's grandfather's brother, it's French House Island, where the brothers were born and grew up together. The map records where they swam and clammed and dug potatoes, as well as the shed where the family salted cod.

Brian is leaning over the table, showing me how an *Ascophyllum* harvester makes a stock assessment of the amount of seaweed growing in his or her chosen sector. It's all about sampling and statistics, he tells me.

"First," he says, "let's draw a rectangle." He draws the rectangle on a blank space in my interview notebook, and he's making sure that I'm following, that he hasn't lost me. "So this rectangle is your harvest sector, and it could be, for instance, a half a mile by two miles and you've got to estimate the standing stock of *Ascophyllum*, which means you have to take random samples or else you will end up with a biased assessment. You begin by making a grid."

He draws a grid within the rectangle.

"There. You have a grid. You have to decide how many samples of *Ascophyllum* to take from it. And that's the sixty-four-million-dollar question. The answer is the more samples you take, the closer you get to the truth of how much is there. But we can't sample the universe. We don't have the money or the time. So we estimate."

We estimate by working from a random numbers table, he says, which is a statistical chart. Then a seaweed harvester would make a copy of the table, pick the numbers he or she needs, and head out to the shore with a quadrat, Brian says, explaining that a quadrat can be a simple wooden square.

"We go out to the seaweed sector. We take the quadrat and we throw it over here and over there, depending on the random numbers table we've chosen, and we cut all the *Ascophyllum* where the quadrat lands, and we weigh it.

"Let's say we estimate that we have fifteen pounds per square foot, and we have ten thousand square feet in our sector. We've got approximately a hundred and fifty thousand pounds of *Ascophyllum*. That's it! And we're going to take seventeen percent of that. We're done!"

How can a harvester manage this without his help? I ask.

He laughs. "There are quicker ways," he says. "But this is how I'd do it. And the rules that are being developed will require something like it for the harvesters—they're heading in this direction."

IN THE SUMMER of 2014, the Department of Marine Resources convened the second group to set aside sensitive wildlife areas that would be off limits to any seaweed harvest. A representative from the Nature Conservancy was a part of this group, as were a seaweed harvester and two wildlife biologists. Brian joined them as an applied scientist who had done scientific work on *Ascophyllum*.

Again, the meetings were open to the public. These were quickly more contentious because the public had had a chance to read up on seaweed harvesting and the efforts to regulate it. Again, citizens came to voice their concerns. Some worried that an expanding harvest might compromise the living they made from other fisheries, such as picking periwinkles. Others spoke to our bedeviled history, how one species after another—with the outstanding and unique exception of the lobster fishery—gets harvested hard and eventually crashes. A number of people spoke of seeing shorebird numbers diminish and wading bird and sea duck numbers drop.

They were tired of it. They wanted the plan to tilt strongly in favor of intertidal and subtidal protections for wildlife and for the wildlife preserves along the coast. It was time, they believed, for the department to do more about saving what we had left and to step back from its efforts to endorse one more industry that takes from the sea.

More troubling, disagreements surfaced within the committee about how to protect species and how and where to regulate the cutting of seaweed when a review of the research doesn't offer any clear answers. After a preliminary plan was argued into place and the meetings drew to a close, I drove up along the coast to the Downeast Institute to pay Brian another visit.

He told me that the meetings of the first working group, based on what scientists presently understand of how *Ascophyllum* grows and is used by the biotic community it creates along the shore, were fun. They were, as much as they could be, based on available science. But the second group, whose task it was to set aside protected areas, was hard for him.

We sat down in a brand-new laboratory classroom, complete with aquariums, pipes that transport seawater, microscopes, and interactive whiteboards. Outside, the tide was rolling in, lifting the blades of *Ascophyllum* and the tight bunches of bladder wrack that dominate the rocks at the water's edge here.

When I asked him about his work in the group to set aside "no harvest" areas, Brian told me how he sees the interface between the world of science and the complexity of wild systems in the Gulf of Maine. He began with the fact on which everyone agrees: the Gulf of Maine is warming. "There are some folks who are saying it may not be warming at as great a rate as we think it is because the way the data were collected varied," he says. "So I think we're all OK with saying yes, it's warming at a fast rate, but we don't know if it's warming at as fast a rate as some say.

"I'm a throwback ecologist to a time when people developed experiments that you can actually manipulate," he continued. "For instance, the density of clams in a mudflat: You put ten in a place or you put seventy in a place and you come back and you've got ten or seventy or you've got fewer. But they haven't swum away. They haven't gotten up and walked away. Ten or seventy aren't there because most likely something came in and took them. And you design your tests for that. It's fairly easy working with a semi- or a completely sedentary species. You can't do that with fish or whales or sharks or lobsters. Those systems are at a level of complexity that goes far beyond anything I could even imagine.

"For instance, trying to understand how to bring back the cod population or trying to understand why we have so many lobsters is fraught with problems. We can talk about cause and effect, but there's only one way you can actually be certain that something causes something else. That's to do an experiment.

"The scientists working out in the Gulf of Maine are focusing on fishing pressure. So what if the loss of fish stocks isn't fishing pressure? What if it's ocean acidification? What if it's warming temperatures? You can't manipulate fish densities and you can't manipulate what the bottom of the Gulf looks like over a large area.

"A true experiment is one that has some controls in it: You don't manipulate something. Then you do. The problem is you can't do that on the scale I'm talking about, so you get a lot of arm waving. It seems to me that concentrating on preventing the harvest of *Ascophyllum* and ignoring all the other threats to the Gulf of Maine is misplaced. What about scallop harvesting? Have you ever seen a video of what draggers do to the bottom? Holy cow!

"But the people around Cobscook Bay, for instance, who have property that touches the tide line, they're focusing on the *Ascophyllum*. And a hundred and fifty feet from where they stand there are thirty scallop boats out there dragging up the bottom. Doesn't that have an impact on the marine system? How come there's no hue and cry about that?

"Something as simple as taking a clam hoe and turning over the mud, well, if you're a bird depending on those clams, and the tide comes in and the clams are not near the surface in the mud, they've gotten turned over and upside down, and they're not available until they get upright again, that's a disturbance. If you choose to ignore these things, then my question is, are you really concerned about the cumulative threats?

"There are things that can be done. I think the reason why we're not doing anything about dragging, for one, is that it's been in place so long. But if you're interested in the environment, I don't think you have the right to pick and choose.

"What I said many times during those meetings is this: What are the disturbances that cause an ecosystem not to be able to rebound or recover? What if the relative threat caused by the *Ascophyllum* harvest could explain one percent of that, and scallop dragging could explain thirty-nine percent? If we are going to make policy about seaweed harvesting, I want to know what the impact of it is compared to the other fisheries.

"Trying to get systems back to what they were is an interesting path. I just don't think we're ever going to be able to do that. It's a great thing to increase the runs of alewives, or to get salmon to come back to our rivers. The problem is in assuming that once everything is in place, that cod, let's say, will return. But we've shifted all the bases: we're increasing water temperatures, we're increasing ocean acidification, and we're increasing the amount of invasive species that are on the bottom. I'm just concerned that if we remove dams, which is fine with me, and we get alewives to run, which is fine with me, I don't think that the equation is $x$ plus $y$ equals $z$.

"I don't think if you bring the alewives back that you're going to get an increase of cod. I just don't think it's that simple. If it were 1930, I'd say that's the equation. If we could take a step back in time and say yes, this is the way to fix it, that's great, but we're not in the 1930s. We're in the 2000s and the whole bottom doesn't look anything like it used to. Yes, the ledges are still there, but there are different organisms all over the place that are taking up habitat.

"It's so complicated! It is *so* complicated!"

❧

*The Question Is Not What You Look At,*
*but What You See*
—Henry David Thoreau

Nancy Sferra and I are sitting across a conference table from each other in a room in the Nature Conservancy office in the town of Brunswick. It's early April. Outside, the Androscoggin River, flooded with winter meltwater, surges over the town's towering hydroelectric dam.

Nancy is a biologist by training and the Maine stewardship and management director for the Nature Conservancy,
an international organization that has saved over one hundred million acres of land and shoreline around the world.
She's been at this work for twenty-one years, and some of
the most valuable land in this state, especially along the
shore and out on the islands, is protected thanks to her work
and her organization.

I've attended many meetings where she was one of the
people making policy, and I am a fan. She speaks clearly and
briefly, she listens carefully, she states what she believes, and
I've never seen her waffle. Like Brian Beal, she sat on both
the Department of Marine Resources' group to make state
policy and on the group to set aside conservation zones.
During those meetings, she defended the concept of the
precautionary principle.

"To me," she explains, leaning back in her chair, "the precautionary principle means, 'Let's not mess it up.' It comes
into play when we really don't know what the impacts of
something like harvesting rockweed are on an ecosystem in
the long term. There isn't a good track record of long-term
research on the results of rockweed harvest. It's hard and
very expensive research to do.

"Suddenly, you're opening up new harvest areas.
Applying the precautionary principle to some of those

places that are, for instance, really good for shorebirds or ducks or whatever—and we know they are because the birds go back to them year after year—means you don't go in and change that. You don't do it, because there's something about that system right there that the birds really like, whether or not you know exactly what it is they like."

More and more, in environmental work, when cause-and-effect relationships have not been proved and there is a realistic possibility of harm, biologists and wildlife managers put in place safeguards based on the precautionary principle until such time as a solid body of research can supply reliable answers. Thus the burden of proof will rest on those who want to advance actions that may cause harm, rather than on those who favor protections.

Nancy mentions the Acadian Seaplants' experimental cutting site that Brian took the working group to see, saying that indeed it didn't look as if it had been cut much at all, and it was cut only once. It was good management, a good way of harvesting. Real life, she notes, doesn't always work that way, and harvesters need to know where other harvesters have been and where they themselves have been so they don't go in and cut a second or a third time. Out in the field, along such a convoluted shoreline, the precision of identifying designated sectors, and what can and cannot be cut, is hard to come by. Unfortunately, she says, the state's marine patrol doesn't have the manpower to monitor these areas. Instead, they investigate when a complaint comes in.

"There would be a lot more comfort in these coastal communities if the work was done by independent, local-based cutters instead of some entity coming in from Canada and hiring seasonal workers," she says. "The harvesters working their own businesses and selling the *Ascophyllum* are most often pretty careful. But still, I'd have liked more in the final plans that was protective in nature.

"Look, we know we have a problem with ocean acidification. We know we have a problem of increased water temperature. We know we have sea-level rise." She's ticking the challenges off on her fingers. "We've got invasive, nonnative species moving in. How will all these things affect *Ascophyllum*? We have no idea. We don't know! We're harvesting it, and we don't know.

"I'm a wildlife biologist. I'm not doing research. I do management. Over the years, I've seen things change. Even if there weren't all this human-caused change, there'd be changes from something else. Nothing's static in the world of science, and we learn to live with uncertainty, and we need to learn to plan for it."

The Nature Conservancy owns spectacular shorelines. The state management plans for *Ascophyllum*, including the protected areas set aside for wildlife, have not addressed the issue of legally conserved shorelines such as theirs and whether those intertidal areas can be open to seaweed harvest. But that will come, because people like Nancy Sferra, working in land trusts all along the coast, will demand it.

"I think both sides can win in this. At TNC, we'd be interested in having a more protective plan in place on our shore lands which says, for instance, you can harvest here this year, but we don't want to see it harvested again for five or ten years. Harvesters may say that three years of no harvest is enough for the resource to rebound, but I'm not sure that's true in every situation. So let's just make sure after a harvest in a particular place that everything grows back and that the species that should be there are there. Then, OK. You can cut again following the harvest plan's guidelines. That's the precautionary principle."

In the UN Code of Conduct for Responsible Fisheries, approved by member nations in 1995, one of the general principles reads like this: "States and subregional and regional fisheries management organizations should apply a precautionary approach widely to conservation, management and exploitation of living aquatic resources in order to protect them and preserve the aquatic environment, taking account of the best scientific evidence available. The absence of adequate scientific information should not be used as a reason for postponing or failing to take measures to conserve target species, associated or dependent species and non-target species and their environment."

As I drove home up the coast after talking to Nancy, I was thinking about this code of conduct and remembering an event in the winter of 2012 that was indicative of how fisheries policy still operates without long-term precautions, and I'm thinking it must have driven both Brian and Nancy nuts. It began with a three-year closure of scallop beds in Blue Hill Bay because of overharvesting. After the hiatus, the Department of Marine Resources policy makers decided it was time to open the bay to scallop fishing for two days every week: one day for draggers, the other for divers. Diving is a completely habitat-safe way to fish for scallops. It is nothing more than picking the shellfish from the bottom. The wear and tear is all on the divers, who are usually young and very fit.

Ours are sea scallops, not the delicate bay scallops that thrive in the warmer waters off Cape Cod. Here, the sea scallop can grow to eight inches in width, with broad, off-white, fan-shaped shells. The tiny spat of the year, after drifting, settle and anchor themselves onto pebbles and seaweeds, grow shells, and become bivalves, two-shelled mollusks, that rest on the bottom. They can swim swiftly for short spurts, snapping their shells like a bellows, powered by the large adductor muscle in their soft, glistening bodies. What we eat when we eat a scallop is that muscle that holds the shells together, the mollusk's locomotive engine.

The scallops that live in water close to shore, such as those in Blue Hill Bay, filter-feed on seaweed detritus. In water offshore, they take in more phytoplankton than seaweed residue because that's what the deeper water supplies.

Morgan Bay, where I live, flows into Blue Hill Bay just beyond Jed Island. These indentations and extensions of salt water are really a single moving body, stretching into shallow inlets, along peninsulas, and around the inshore islands where the seaweeds grow, and in that particular winter that I remember, we who live here were aware that the first scallop-fishing day, a Tuesday, would be offered to the draggers. Many of us also knew that draggers often take a first run along the bottom of a bay to sweep the area clean of seaweeds, mainly kelps, before they turn around to go back for the scallops.

From close-by harbors and from farther down east, fishermen arrived with their boats rigged for scalloping, idling just outside the demarcation zone that was soon thrown open to them: our own Oklahoma land rush. By sundown, the beds of the outer harbor had been raked clean. In disgust, the DMR commissioner called it "a derby mentality" and shut the bay down. Again.

Was anyone really surprised?

I attended a memorial for a neighbor a few months after that, under a big white tent by the shore, and found myself talking to a diver. When I asked him about the scallop dragging, he looked nervous and said we should speak outside the tent. We went out and stood apart from other people, most of whom were in the fishing business.

"It's bad," he told me quietly. "Sometimes the bottom is so chewed up, I get out of the water and just go home, but if I said that out loud, I'd be in trouble."

## Every Cog and Wheel

Elinor Ostrom, a Nobel Prize winner in economics, spent most of her life studying how individuals in communities with a successful and robust wild commons work together to preserve and harvest what they own. Her invaluable study is slowly gaining attention in the world of wild resource management, and the key to her vision of a functioning commons depends on a cohesive community bound to a home place.

Ostrom starts with borders. A community owns what? How far, how deep? She moves from the measurement of a place to the importance of understanding local needs and conditions. How much seaweed, for example, or how many scallops can a community's individuals take from within its boundaries without harm to the environmental systems of the surrounding water? Without eventually starving itself to death?

If a community directly participates in rule making, she argues, as the seaweed community seems to have done at the Department of Marine Resources meetings here in Maine, it is more likely to stand by those rules and enforce them within its own ranks. The rules to protect and use a commons must be honored by those outside it, Ostrom writes, with the legal understanding that a commons belongs exclusively to that community. And finally, the community must figure out how to monitor its own members' behavior. Infractions are a fact of life. How will it sanction its violators?

This is new science based on an old human endeavor to keep a group of people thriving within its essential but limited wildlife supply, and as history has proved, it's an endeavor honored mainly in the breach.

On that long drive home from the Androscoggin, I recalled another story, one that illustrates a preliminary understanding of what a commons can become: A few weeks before, licensed clam diggers in the coastal town of Gouldsboro had agreed among one another that they should be taking fewer large "breeder" clams. They met with the town Shellfish Committee to write up a new ordinance for the town that allows every clam digger only 10 percent of a day's harvest of clams over four inches in length, because every female four-inch-long clam that continues to live in a mudflat releases millions and millions of spat. And without a microscope, you can't tell a male from a female clam. Sherman Merchant Sr., a longtime clam digger and a good neighbor of ours when we lived in the town, declared, "If you don't leave some of those big clams, you won't have a future."

He's right. Before the end of the meeting, they hired a local carpenter to make green crab traps, in an effort to rid the bays in town of some of the clam's most efficient predators. This spring, the same group of clam diggers, working with the shellfish warden in town, purchased thirty thousand seed clams from the Downeast Institute on Wass Island, planted them in the mudflats around town, and covered them with plastic netting to keep away green crabs.

Not only is this group far exceeding the state requirement for clam diggers to work a few hours a year in conservation efforts, but they are healing the excesses of the present to protect the future. They know their bays, and they are stepping forward to own them.

Ostrom would applaud these first steps toward taking responsibility for working in and conserving a shared resource, especially because these steps were not imposed from the top but generated by the people who do the work.

I can't help wondering whether the attention brought to the rockweed harvest, and to the fact that the harvesters and processors asked for regulations to oversee their own industry, was noticed and appreciated by the citizens of Gouldsboro.

Because that's the way change begins.

ON AN AFTERNOON in early fall, I was exploring seaweeds in a tide pool in the town of Milbridge along a rough line of coast. The water was dead low and strands of *Ascophyllum* hung thickly from their holdfasts down the slabs of granite above me. No one else was along that part of the shore, and when I looked up from my small vantage, the layered strands stretched under the overcast sky beyond the curve of the bay. *Ascophyllum* from here to the end of the world. And before me, the sea, kicked up by a wind, went on beyond the islands—forever. I stood for a moment in a state of mild shock, as if I were somehow a castaway. I was no more a presence there than a small green crab scuttling along the edge of the tide pool. How could I damage any of this? But of course I could, and we have. And so, by the way, has that crab.

All of us have had moments when the wild world overwhelms us, as it must have done at times to the Abenakis and to the first Europeans who stepped onto this complicated shore. We have our stories of being little in big landscapes, although we know that the resources of the earth are finite. We know this. Or do we?

When sitting in on a meeting of seaweed harvesters and processors, I heard one of the participants, tilting back in his chair, declare, "Rockweed—you just can't overharvest it. You can't. It's infinite!" His right hand went up and flipped away any suspicion to the contrary.

I watched his hand because it looked like the wing of a bird to me, and it sent me, for a dreamlike moment, to the edge of a scrub field somewhere in Virginia at dusk in the late 1800s as clouds of birds—hundreds of the now gone, fabled passenger pigeons—settled into their nighttime roosts in a copse of live oaks. Across the nearby fields I could almost hear the echoing voices of the gunners as they came running with their lanky dogs.

How we adjust to the fact of limited supply and the complexity of wild systems should guide what and how much we take and what we leave alone. It's the base from which we build an ethic for both land and water. But our behaviors are still driven by vestiges of outdated beliefs. What's required here is not only a behavioral shift but a deeper one that reconfigures how we humans see our place in the world and, equally, how we value the lives of species other than our own.

In 1968, Garrett Hardin wrote a seminal and compelling paper titled "The Tragedy of the Commons"—later he said he should have called it "The Tragedy of the Unmanaged Commons"—in which he argued that we, as individuals, routinely act in ways that are destructive to both human and natural communities, and that maintaining a commons is undermined by an individual's pursuit of more than what a shared commons can provide.

How can we disagree with this assessment? For those of us who have spent our lives in coastal communities, it rings true. We've seen it out on the clam flats, in the Atlantic herring fishery, and in dragging for scallops and sea urchins. And that's the short list.

Carl Safina, marine biologist, award-winning writer, and founder of Blue Ocean Institute, has argued, like Hardin before him, that we carry within us values that reflect a time when we knew little about the world through science, and more about what it was and who we believed we were through myth and magic. And in myth and magic we are at the center of endless supply.

People who work to conserve or preserve or repair wild systems believe that they are working against the clock. Some insist that the only way to protect what we have from the onslaught of climate change is to put aside large areas free from all harvest, because they will be where many species native to our waters will find refuge and can start, if they are able, to adapt.

PAUL MOLYNEAUX TELLS me that he used to be able to hear the tide turn. "It slaps the rocks a little differently when it starts coming back," he says. A commercial fisherman in southern New England for two decades, Paul built boats for a while and managed a fish-processing plant on the Passamaquoddy reservation at Sipayik for a few years. Then he built himself a dory, bought himself a pair of oars, and became an independent fisherman, an owner-operator. He's energetic and muscular, a man who loves the sea and its wildness.

"I made my living with a rowboat. There was still a place on the coast where I could make my living with a rowboat," he says, and then he laughs, incredulous.

We're sitting on a log, a worn item of flotsam that has come ashore at Haycock Harbor, about forty-five minutes from the Canadian border. He lives in an old farmhouse not far from here with his wife, Regina, and their two children.

It's late fall. The tide is out, the sky uniformly gray, and the narrow harbor, like a wedge, ends in a dock for a single lobster boat. Below us, on the cobbles, a woman is picking periwinkles, and across the way, seaweeds cling to a ledge, layered in horizontal bands according to species: spiral bladder wrack at the top, then *Ascophyllum*, then northern bladder wrack, and below that, purple laver.

We had just hiked up a short promontory to get a view of Grand Manan Island, Canada, ten miles offshore. Paul rowed his dory to that island with a friend some years ago. It took seven hours through a white-capped fetch, a crazy venture that he's quite proud of. These days he spends his time writing for *Fisherman's Voice*, a Maine publication, as well as for other newspapers and magazines, and he has three books to his name, two of them on fishing and fisheries policy. In 2007 he won a Guggenheim Fellowship for this work. What interests him is writing about how fisheries policy affects the independent coastal fishermen and fisherwomen and the environmental integrity of the oceans, and from the beginning he has been involved with issues of seaweed harvesting and regulation.

"I'd come down here in the dark," he tells me. "It was just turning light when I rowed my dory out with the tide. I was picking periwinkles and hauling lobster traps and harvesting rockweed that I sold to people for their gardens. I'd go out with the tide, work through low tide, and come back with the incoming tide. Sometimes I'd row eight miles a day. It was a beautiful life. It was meditative."

When I asked him what changed, he said, "I didn't want to get big in the lobster business. I was hauling fifty traps by hand, and I really couldn't afford to pay for a license anymore. I had to get big or get out. So I got out.

"There were no regulations on periwinkle picking then, and some harvesters from the next town came in here one summer, working the places I had already picked, and they kind of put me out of business. I used to leave most of the periwinkles, but they took every last one. Little ones, big ones.

"Then the urchin fishery opened up," Paul says. "Everyone was diving for them. I went urchin diving. I loved it and so I fought tooth and nail to preserve what we had through tote limits and by scaling back the harvest. But what I said didn't make much difference."

During the green sea urchin boom, the urchins were sold to Japan, where the bright orange roe is considered a delicacy. In what seemed no time at all—from the mideighties to the midnineties—sea urchins were all but cleaned out, and another fishery fell away. Today, more safeguards have been put in place, and the urchins are back, although they have never fully recovered.

Both sea urchins and periwinkles are grazers. They depend on a variety of microalgae and seaweeds for their survival. You have two vulnerable but still functioning fisheries, and then you insert a third one: their food source. That's the one you have to watch, Paul tells me.

When the Canadian company Acadian Seaplants came down into Maine to begin harvesting *Ascophyllum*, he says he thought it was nuts to let them in, although we have no laws to stop them.

"It's such an important habitat, and to allow a foreign company to cut it and pay our people pennies a pound instead of doing it ourselves sustainably . . . !" He shakes his head. "In exchange for handing over our increasingly valuable rockweed to a company that will ship it by the ton to Japan, Singapore, Brazil, and numerous other countries, Washington County gets a few low-paying jobs for harvesters, many of whom don't even live here year-round!

"We don't know how to assess the value of species within their ecological communities," he says, "so we tend to think of them as worthless rather than priceless."

# CHAPTER 10

# The Common Eider Duck

In Yellowstone the work to restore gray wolves went on for years. In the North Cascades in Washington State, biologists have drawn up a plan to protect a last population of grizzlies. Here on this coast we've spent time and money to monitor the movements of humpback and right whales, to disentangle them from lobster trap gear, and to protect piping plovers, Atlantic puffins, roseate terns, and more. We know what loss means when a species native to its home place loses ground. For many of us, it's a long, sad echo in the heart. Some biologists spend their lives working to stabilize populations of native species that have evolved within a place and with other species. They do it because we want our places whole, deeply layered, and alive.

"Eider ducks are the poster child for rockweed," Brad Allen tells me. "The ducklings need to be in it within hours of hatching to glean the invertebrates; they need that vegetative structure at the surface for the first couple of weeks of their lives."

Brad is a biologist and the bird group leader for the Maine Department of Inland Fisheries and Wildlife. He and Lucas Savoy of the Biodiversity Research Institute, a Maine-based organization with a worldwide reach that assesses emerging threats to wildlife and ecosystems, have placed small radio transmitters surgically onto the backs of fifty hen eiders, and now, as the summer wears on, they are tracking them off the Midcoast mainland and around the islands, counting the ducklings.

It's a late morning in June. Lucas has left Brad and me off on the shore of an island and gone on in his outboard to count birds, and I am following Brad up over the rocks of a small low-tide cove, covered in densely growing *Ascophyllum*, and into the pathless interior of this twenty-seven-acre eider-nesting island in Casco Bay. It supports about 250 eider nests a year and was purchased by state land trusts with federal grants and oil compensation funds. It's managed by MDIFW—that's Brad and his crew. These beautiful ducks are closely associated with *Ascophyllum* at pivotal stages of their lives. With the increased cutting and harvesting of this species of seaweed, we have no clear understanding yet of how or whether the industry will affect them and their population, which already struggles along the coast.

The public, a nesting eider's nightmare, is not allowed on this island from April 1 through the end of August, which guarantees the birds the privacy they need. They share the island, however, with a number of other nesting species, including glossy ibises, snowy egrets, black-crowned night herons, great blue herons, and a pair of bald eagles.

Common eiders are North America's biggest ducks, strong and very beautiful with their elegant plumage and signature Roman profiles. They ride the water out in our bays and in the heavy chop around headlands, coming and going throughout the year.

When I lived in Gouldsboro, years ago, I used to find them in the winter down at the harbor on the coldest mornings, the sea smoke in thick drifts over the water, and when those gauzy curtains parted, the eiders appeared, resting between the fishing boats at their moorings, quiet in the stillness of those incomparable days. The curtains would draw closed again, but I knew the ducks were in front of me, a dozen yards away in the bone-cold water, though all I could see was sea smoke and my own breath in the air.

"I love these ducks," Brad tells me as we creep along into the woods. His job is to keep their population steady and to figure out where the biggest pressure points are and what they can't live without. Mature female eiders' annual survival rate along this coast is at about 90 percent, he tells me. The males survive at about 84 percent. Those losses are due, he believes, to disease, starvation, hunting, and the birds' being taken by the growing eagle population and the black-backed gulls. The percentages are fairly constant. An eider is a long-lived duck—twenty-two years is the known record—and a drop of 10 to 16 percent of the population every year indicates a relatively stable pattern, but only if the young grow up to replace those numbers. On this coast there is a radical downward population trajectory, and Brad thinks that the normal number of young that perish each year—although stunningly high at 90 percent—has probably risen to, he guesses, 95 percent.

That number is too steep.

"Why might it be about ninety-five percent? Because gulls and eagles may have nothing else to eat. There's not enough food out there for everybody—including eiders. Gull predation on eider ducklings is much higher now, and that's, I think, a major cause of our population decline.

"When the alewife run coincides with the eider duckling hatch and every black-backed gull on the coast is more interested in eating alewives than ducklings, that's a buffer. In addition, if we recover other fisheries, then perhaps eagles will eat fish again, instead of eider ducks, and great blue herons and gulls will eat alewives instead of ducklings." Which leads him to the subject of the rockweed harvest plan developed by the Maine Department of Marine Resources and the meetings to set aside protective areas of no harvest. He thinks the *Ascophyllum* harvest may put pressure on ducklings, already under pressure, and any more could cause catastrophic decline.

"Sometimes I think it's the fox guarding the henhouse," he says, referring to the tendency he sees of the Maine Department of Marine Resources people to put the seaweed harvesters above the work of resource protection.

"We made recommendations. We were asked for a list of rockweed harvest areas to close for the protection of eiders. What we gave them to protect was such a small percentage of the coast of Maine and the rockweed, and we recommended seasonal restrictions on rockweed harvesting on seabird nesting islands when the birds were there in the spring and summer. And the rest of the year, we said OK to harvest. But they beat us back on almost every point.

"Frankly, I don't need a scientific paper from somewhere like British Columbia that says rockweed is important to these birds. We know it is."

However, he tells me, his department has started to work with researchers at the University of Maine to study invertebrate numbers in both harvested and unharvested rockweed areas.

"The rockweed harvesters tell us they don't take bycatch, that they throw all the creatures that come up with the seaweed back. We're planning our own study to find out if that's so."

I REMEMBER AN afternoon in March of this year, the sun dropping beyond the Salt Pond to the west, and I was leaning into the guardrail by the South Blue Hill Bridge as traffic rattled onto the metal deck. Ahead, to the east, a raft of perhaps eighty eider ducks bobbed together in the swirl of tide. The water was building swiftly over the Irish mosses and the tufted redweeds and the beds of blue mussels stacked in crevices between the rocks.

This is the eiders' favorite tide, the water racing through the embankments, a swift, clean recharge, perfect for their favorite food, the blue mussel, for the stirring up of the microscopic plankton and seaweed detritus the mussels feed on, and for the ducks to dive down to get them.

Above the noise of the traffic and the tide, I could make out the *oo-woo-oo* voices of the drakes, a low-pitched, incessant moan, like a group complaint. But of course it's not a complaint at all. It's a song of springtime and delight.

Eiders are tough. Apocryphal stories about them abound. One, found in old birding texts, recounts that some of these ducks, when shot by hunters and mortally injured, refuse to be taken, and dive, clinging to underwater seaweeds with their bills until they drown. Biologists don't subscribe to stories such as these, but they do say that the birds, when injured by hunters, will often dive, come up somewhere else very quietly, and slip away low in the water rather than show themselves.

BRAD AND I are creeping along Indian-style, talking in whispers, both of us wearing bright white hazmat suits with hoods, and we've pulled on booties and gloves. We look like two giant, puffy snowballs in the mid-June heat. The reason for this getup is all around us: poison ivy snakes through the undergrowth and coils along anything that stands upright. It also seems to have managed to transform itself into bushes and small trees, its bright, noxious leaves shimmering in the morning sun.

Brad is a stalwart man, nimble on his feet, a day's worth of pepper-and-salt stubble on his cheeks. He's got a quiet voice and a forthright manner. I tiptoe behind him, listening as he turns to tell me something more about eiders and wildlife policy, my attention only somewhat diverted from what he is saying by the gleaming green poison all around me.

Trunks of fallen spruce and ash and oak litter the island floor, and under them, when there is space enough for a female duck in the deep, soft beds of dead leaves, we find their favorite nesting spots. Brad carries a long-handled net because he intends to catch a female and put her in my hands so that I can understand by holding her, he says, the sharp edge these eiders live on. It allows for very little to go wrong, because too much of one thing or not enough of another can cause immediate harm. Tough ducks on a fine edge.

When the male eiders vigilantly guarded the females as they fed this past spring, there was good reason for it. By the time she leaves the water to seek a nest site, the hen is heavier than the male and ready to spend twenty-six days incubating her eggs—usually four of them in a clutch— with few breaks. And her eggs, a smooth gray-blue-green without much sheen, are large, nearly twice the size of a domestic hen's egg. So what she ate, and how much of it, are crucial both to her having the substance within her to make those eggs and to her being able to sit tight for almost a month's fasting and incubation.

Brad lowers his net, quickly dropping it over a hen nesting just under the trunk of a fallen spruce. There's a brief scuffle as he lifts her, and then she goes quiet, and he hands her to me. I pull up the sleeves of my hazmat suit, and she rests in my arms. She's unmoving, her brood patch—the featherless spot on her belly with which she warms her eggs—hot against my left forearm. And my thought is, right then, that we all need to feel the hot touch of a wild life against our skin like this.

Brad and I are near the center of the island, in its most forested site. Astonishingly, two eagles tend their nest above us. They won't come down through the trees. They prefer to hunt around the island's edges and across open water.

"Feel how thin she is," Brad says quietly. And he's right, she's almost as light as air, and it's little more than halfway through her nesting cycle. I hold this individual life, thin, probably hungry, and doing what her species has done forever. By the time she leaves here with her ducklings in tow, she'll have lost a quarter of her body weight. That she will have enough energy to swim, feed, and protect her young astonishes me.

I hand her back to Brad. He sets her down on her eggs, and we leave.

"The ducklings will hatch into that little bowl of down the hen's made," Brad says, "nature's best insulating material in the world."

It can take as long as a full day for ducklings to break free of their eggs. A clutch will hatch in synchrony, not only because the hen began brooding the eggs all at one time, but because the hatchlings respond to one another's voices even as they work to crack their shells and peel their way out. Then they fluff and dry in the comfort of their mother's down, nuzzling their own new down and making tiny high-pitched sounds. The ducklings are born with a small egg sack inside them, which will nourish them as they dry off and start their walk, single file behind their dam, across, in this case, the island, down the carpet of *Ascophyllum* in the cove, and into the water. Within a short time the hen will take them to the mainland, where there is more food along an easily accessible and extended coastline, with variety for both her and her ducklings.

From this nesting island to the mainland, it's a mile and a half across open water. The ducklings—each one of them you could cup in the palm of your hand—follow the hen in a line across that long and shifting expanse, paddling vigorously, brown balls of fluff too buoyant to dive.

A black-backed gull or a bald eagle may try, says Brad, to pick them off "and eat them like popcorn." But hens will muster an energetic defense of their ducklings. Still, it's amazing that any ducklings survive this long trek. But they do.

When Lucas swings back in his motorboat and picks us up, we head to the mainland, to the tidal edges where eider hens and their young congregate in the seaweed beds.

Brad and Lucas are listening to the pulse of beeps that they have picked up with the antenna on their tracking machine. The beeps tell us that a hen with a radio transmitter is nearby, and then we see her with two other hens and at least a dozen ducklings against the mainland shore in a high-water patch of *Ascophyllum*, the young twirling and bobbing in it, feeding off the fronds that hold periwinkles, tiny mussels, little tubeworms, young whelks, scuds, isopods, and more.

What these ducklings need in their infancy is abundant and undisturbed *Ascophyllum* that grows tall enough so that even in a full tide it stretches up and then out horizontally across the water, like a raft. If the fronds are long and thickly branched, the young rest on them and feed from them throughout the tide cycle. As the ducklings grow and become stronger, they begin to make shallow dives, taking the small animals they need from deeper levels of the seaweeds, and eventually they dive to the bottom, and their feeding area and prey base become larger and more plentiful.

Lucas counts the radioed hen, the two others, and the crèche—or group—of young, adding this information to what they already know about the hen they are tracking, building a biography of sorts. Even though hens may gather their ducklings together to form a crèche and share the duties of protection and finding areas where their young can feed, over time a narrative does develop, and it gives Brad and Lucas a good sense of how many ducklings survive in a year.

We sit in the boat, the motor turned off, and watch how the ducklings before us seem to vanish into a bed of *Ascophyllum* as it gently moves to the water's rhythm. This remarkable, regenerative, abundant species of seaweed offers them a good start to a long life, and it matches them in color and pattern so perfectly that a marauding gull just might pass them by.

"I'm looking for things I can improve to help increase the number of these birds that survive," Brad tells me quietly as we watch. "I'm guessing duckling survival is the main problem. If it turns out to be, then I'll have to figure out how to deal with it.

"One step at a time," he adds, a bit wistfully. This is the first year of the three-year study to see how many young make it through the crucial period when they march behind the hens from their nests to the water and learn to feed and, eventually, to fly.

Up ahead, in this cove ringed with *Ascophyllum*, we find three wide-mesh bags of the seaweed floating midwater, as big around as VW bugs. They drift slightly, all cut and packed by the man who now stands on the deck of his mechanical seaweed harvesting boat, smoking a cigarette and taking a break.

He waves at us, and we wave back.

# CHAPTER 11

# The European Green Crab

Seaweeds are good for many native marine species, of that there is no doubt, but sometimes invasive species that can overcome wild native populations and change the dynamic of the shore use these same seaweed habitats too well. I hold in my hand a European green crab carapace—the top shell—which I picked up at the bay yesterday. It was lying on windrows of seaweeds pulled loose and thrown ashore by the high tides and summer storms. The shell has baked in the sun and turned a soft orange color. Five sharp incurved points run along each edge to the eye sockets. The eyestalk stems are neatly folded into the sockets, and between them a rostrum, like a narrow visor on a cap, projects in three lobes. This carapace was shed, and the new, larger crab now meanders somewhere out in the bay—hungry, intrepid and, so far, invincible.

We know the stories of how our cosmopolitan lives can fracture a delicate balance that native species have made with one another and with the places where they live. We move around a lot, trailing species from one location to another. Between climate change, extinctions, and international trade, we erase or imperil thousands of years of evolutionary adaptation and set off new bouts of survival of the fittest. One of the fittest invaders we've brought along with us is this crab, *Carcinus maenas*.

Seaweeds, especially *Ascophyllum* and the *Fucus* species, play a role in its success. When the floating green crab larva first falls out of the water column, sinks to the bottom, and takes on the shape of a tiny, translucent new crab, it can be gobbled up by almost any predator. Its safety lies in the long, sheltering fronds of the rockweeds, where it nudges itself into the crevices of the rocks and is draped within the protective seaweed blades. The crab's enablers, they offer it shelter and food.

Here on the Maine coast, the green crab is *the* inshore crab. Two endemic species, the rock crab and the Jonah crab, are larger and live in deeper water, both of them regularly caught and sold for food. The great spider crab, *Hyas araneus*, also endemic, lives in between the range of the European green grab and the rock and Jonah crabs. And then there's the little hermit crab, tugging around a snail shell, its borrowed home.

Masters of the intertidal zone, the young green crabs keep close to the safety of the nearby seaweeds, but as they grow they become bold predators, dancing and digging their way over and into mud, across sand and marsh and cobble—always hungry. This isn't a big crab, but it has had an out-size impact, diminishing our abundance of native species. It has changed this big, complicated coastline, marginalizing the lives of coastal people who make a living digging clams and bait worms, harvesting mussels, and fishing for scallops. Until we figure out how to fight it, or manage it, or decide how we'd like to cook and eat it, we are stuck with it.

The first published report of a new species of crab in the Massachusetts intertidal zone was written in 1817, but the crabs had probably arrived years before. This was the era of sleek wooden trading ships racing west across the Atlantic to pick up American white pine and beaver pelts, rum and salt cod, corn and block ice packed in sawdust. The stowaways most likely embarked somewhere along the British coast. At that time, ships that were not carrying cargo, crossing to buy rather than to sell, used dry ballast to steady the hulls and to counteract the weight of the masts, which made the ships top-heavy. Dry ballast would have been stones.

I imagine sailors somewhere at a British port pulling stones off the shore and out of the upper tide, carrying them up a gangplank and into a ship's hold. Perhaps a number of the stones had seaweeds growing on them, and within those wet handfuls of algae crouched a young green crab or two. While sheltering in their home cove, they were lifted and taken away. Green crabs can live two months out of water as long as they have enough moisture to keep their gills wet, and the holds of those ships were perfect redoubts for a crab's transatlantic voyage: damp, briny, and dark.

The native range of the green crab runs from the coast of Norway, down the eastern Atlantic, out to the British Isles, and south to North Africa, where it lives in competition with a number of other species of inshore crab. It possesses an array of talents that make it a gifted adapter to new environments. One of these is high fecundity. The females can brood and release 180,000 eggs, sometimes twice a year. A prolonged larval stage means it can also float for days in the ballast water carried in metal-hulled ships, which quickly became far more efficient crab transports than the old wooden-hulled trading ships. It eats voraciously, omnivorously, and it can tolerate a range of salinity and cold.

By the late 1940s the European green crab was a fixture in the Gulf of Maine, its numbers waxing and waning in response to the severity and length of the winters. It spread south and has been halted along the Maryland and Virginia coasts, probably by the presence of blue crabs.

ON A DAY in early July, out on Wass Island, I accompanied Brian Beal as he drove his truck down a dirt road to a bluff above Molly Cove. We had come to visit a project he has organized, with the help of the Beals Shellfish Conservation Committee, for students in the seventh and eighth grades of the local elementary school.

"We're teaching kids how important the resources are here. They grow up in families harvesting lobsters and clams, but where those clams come from or where those lobsters come from and how they live aren't usually conversations within a family. It's just, 'How many did you get today?'"

Not only is he teaching his regular classes at the University of Maine at Machias, but the Maine Economic Improvement Fund granted him $200,000 to continue his soft-shell clam research out on the flats in coastal towns around Casco Bay, focusing on bringing back clam populations and stabilizing that fishery for the local diggers. This means that Brian is spending a lot of time thinking about green crabs and how to minimize their impact.

He's invented "Beal boxes," devices to prevent crab predation on shellfish. They are used by students in high school science classes to study the relationship between the crabs and the clams on their neighborhood clam flats. And here, at the Downeast Institute on the island, he teaches kids about using the scientific method to help protect the marine environment around them, the environment he hopes they will inherit someday and care for.

As we drive, I ask him about shellfish and seaweeds—whether, for instance, clams can siphon seaweed detritus into their bodies and use it as food.

"Clams are filter feeders," he explains. "They have a certain size particle they can ingest. Most of it comes in the form of microalgae—phytoplankton—in the range of five microns to a hundred microns. Very, very small. Mussels have the ability to ingest larger particles, so it is possible that seaweeds can contribute directly to the diet of mussels through detritus.

"When seaweeds break down and die, they are sloughing off and coming apart into smaller and smaller particles, and those nutrients contained in them are being put back into the ocean, and they help the phytoplankton to grow. It's all a part of recycling. The macroalgae (the seaweeds) break down to feed the microalgae (the phytoplankton), which feeds the filter feeders like clams and mussels."

Molly Cove, a cup of mud-and-cobble shore, looks out across Eastern Bay. It is bracketed by ledges from which hang wet sheaves of *Ascophyllum* and bladder wrack, and behind the ledges lift groves of spruce. As we get out of the truck and walk down to the low-tide line, Brian tells me of an experiment he designed for a class of elementary school children to discover whether green crabs of a markedly small size were able to eat soft-shell clams without breaking their shells, by slipping their little pincers between the two shells and attacking live tissue.

Each student was given a Tupperware container with a screen cutout in it for water circulation. Inside they placed a young crab and a clam, and then they dropped the containers into fifty-gallon tanks. When they returned for the next class and pulled up their Tupperware cages, they found that the crabs, even those no larger than a Jacob's cattle bean, had managed to pluck between the shells of the clams and eat them clean. This experiment proves that crabs, even small ones, are efficient predators of bivalves—two-shelled mollusks—but it doesn't prove that green crabs can dramatically alter a mudflat, eating their way through masses of clams and worms, mussels, and snails, until what is left is just unfiltered mud.

"People make statements that they think are factual, but we need to do the critical tests. A lot of silliness is still being described as science. To be wrong about something isn't something you should shun, it just means you haven't approached the problem correctly. We start with an observation and we make a prediction. That is a hypothesis. There's no way to prove a hypothesis. You can only disprove it. So you have to do everything in your power to disprove what you think is true."

Ahead of us, lying in the low tide, bobbing in the mild chop of water no more than a few inches deep, are four squares of netting, fourteen feet by fourteen feet. They are made of polypropylene, and each is securely tucked into the mud at its edges like a well-made bed. Five toggles are set inside each net, lifting them above the sediment and keeping the nets aloft so that the soft-shell clams beneath them can raise their siphons through the mud and into the water to feed on the microalgae the water brings shoreward during a rising tide. The clams are protected from green crab predation by the nets, and the nets are kept free from the growth of algae by the periwinkles that slide over the tops of them, feeding.

"Most years there are few wild clams in this cove," Brian tells me, "but there are always some nice big ones in the mud." He begins digging with his hoe, which he carried down from the truck, and uncovers a big, fat clam. "Look at that!" he exclaims. "That's a great clam! I'll put it right back."

As we walk in the tide, he tells me to look across the beach and points out what he calls pockmarks in the wet mud—round indentations the circumference of tether balls. They cover the intertidal shore. I assumed a clammer had been down with a hoe a tide or two ago, and this was what was left of his or her digging.

"They're made by green crabs," Brian says.

"What were the crabs doing down in the mud?" I ask.

"Clamming!"

Now as he digs, his hand latches onto something deep and he yanks it up. It's a large male green crab, pincers flailing. All the proof he needs.

Green crabs can move up and down in the mud, Brian explains, but they can't move sideways in it, so the clams that have been seeded under the nets here are safe, unless the nets happen to be placed over a crab already in the mud that comes up into them. That hasn't happened, and the elementary school students have planted three thousand hatchery-raised juvenile clams under each net. In November they'll be back for the harvest.

When I think of the range of work Brian has taken on, stretching up and down the coast, engaging adults, other scientists, and children in the issues of the inshore waters, and also raising seed clams and mussels and oysters and surf clams back at the institute to revive the coastal life of the state, I think of him as a sort of *Catcher in the Rye* character, trying to keep others from running off a cliff, and the cliff in question is this diminished coast that used to be, within his lifetime, a rough but giving Eden.

No one involved in digging clams for a living during the past couple of decades would argue with the statement that the rate of harvest was unsustainable, despite the fact that they themselves were doing it. But today, there's a change of heart in some people who make their livelihoods out on the flats, and they are becoming dedicated stewards. That seems to be a common human response to wild resource collapse. We need to get knocked off our horse before we begin to see the light.

For the past four years, Brian has been driving to the Midcoast, conducting his experiments with help from Friends of Casco Bay and the Maine Clammers Association's volunteers to reestablish the former abundance of their clam flats. Not only is research expensive, but it takes time. It takes patience. The years pile up, and careful science begins to accumulate as the clammers wait for a solution to their problem.

An abundance of soft-shell clam larvae circulates in the Casco Bay water column, Brian explains. Something happens, however, when they settle down on the flats to form shells.

"To think of the way clams come into a flat," Brian begins, "you need to think of holding a saltshaker. Clams are swimming in the water as larvae. When they get to be about two or three weeks old, their developing shell is heavy, and their swimming organ can't keep up, so they settle. I call it the saltshaker effect. Larvae falling out of the water column are just like salt crystals coming out of a saltshaker. They're landing here and there and there. They settle. A stream of millions of clams. A clam that settles is one-fifth of a millimeter, one hundred-twenty-fifth of an inch. They can go right through the holes in the nets we've put down and settle on the mud beneath them."

What Brian did in these towns was to set up experiments that tested two hypotheses: whether the mudflats had such dramatically low pH numbers, owing to the effects of acid rain, that the clams were unable to make their calcium-dependent shells, and whether green crabs were responsible for the lack of young clams growing into harvestable adults. After three years of work and a variety of testing methods, such as putting crushed clamshells onto the flats to introduce calcium to boost the pH numbers, what he has found is that the pH in the bays where his experiments took place had nothing to do with the loss of clams. Green crab predation did.

He set up test plots over and over, because, he says, "Nothing is definitive unless you've done it multiple times in multiple places. And there were always more clams under the nets. It was a pattern. The crushed clamshells didn't matter.

"If you're a community and you have the option of putting out crushed shell on a flat or not, don't bother. It's not going to do you any good. Rather, if you want to do something, you should put down netting. Netting doesn't guarantee you'll get clams—the saltshaker effect, again. But it does say that you'll deter predators."

BACK AT THE institute, we walk to the shore of Black Duck Cove, and I ask what this shore would look like now if it had never met a single green crab. Most of the damage, of course, is underwater, but the crabs have eaten the green seaweeds along the shore, Brian explains, particularly the sea lettuce, *Ulva lactuca*, and there is none left here.

Only the brown seaweeds, the *Ascophyllum* and the *Fucus* species, fringe the land in a slow incoming tide. No doubt within them lithe crabs are moving, preying on almost everything they find, including crabs smaller than they are.

ALTHOUGH PEOPLE WHO have tried green crabs say that their meat is sweet and they make a nice broth, the trouble with eating them is that they are small, their claws narrow, and picking them out claw by claw—which can be fun with Jonah and rock crabs—is an oversize amount of work for a tiny dab of meat. Another problem is that the soft-shell crab market, which works so well for blue crabs, can't work for these: they don't all shed at one time on this coast, and there are no visual clues that we know of to predict when a single crab is about to creep out of its shell to start a new one.

Companies that make compost would be glad to incorporate them into their products, but they want the crabs to be dead, dried, and delivered, and that is expensive work for what they'd pay. Furthermore, it would be difficult to develop a profitable market when populations of the crabs fluctuate, although as the Gulf of Maine warms, the numbers are bound to stabilize and increase.

A friend of mine, a farmer from the town of Gouldsboro, was told by the town's clam warden that the clammers had caught a lot of green crabs in their traps out on the flats and would be happy to bring them over to the farm to use as compost.

He was delighted. "Sounds great!" he said.

"One more thing," the warden said. "They run every which way out of the traps . . . and they're hard to kill."

"Well, maybe not," my friend told him.

What remains the biggest setback to developing a green crab fishery of any sort is that the people who harvest native species of shellfish, snails, and worms, alarmed by the toll taken by these voracious invaders, would prefer a program of eradication. And you can't blame them. If a green crab industry were established, we'd have to protect the crabs, the creatures that have flummoxed the greatest predator of all.

IN SUMMER, NEIGHBORS and I walk down to the bay we live on for a quick swim. The ledge that slants into the high tide offers a steep angle by the road. A few green crabs move diagonally across it. We step into the water, and they approach. Some are large, with four-inch carapace widths, and red backs rather than green. They are the alpha males that have grown thicker shells and larger claws than the green-colored males have, and they are armed to fight for access to females. After a brief toe check, they dance away.

Our bay has changed. We've lost mud snails and blue mussels and soft-shell clams. We've lost our eelgrass beds. Along the head of the bay, to the east of this small swimming cove, the *Spartina* grasses have almost disappeared. The crabs dig caves into what's left of them and feed on their nutritious rhizomes. Chunks of dislodged *Spartina* banks riddled with crab burrows lie upturned against the cobble-and-gravel shore. These were once a gentle, grassy extension of the shore itself, softening the impact of incoming storm surges and high spring tides.

Mature green crabs head out into deeper water in winter to escape the cold. Here at the shore, the young crabs pack together in their sod *Spartina* caves until spring, fifty or more crammed together. Come March, the constellation of Cancer, the sign of the crab, rises directly to the south of this bay, a small group of faraway stars barely visible to the human eye. It shines dimly down on the snow crust and through the occasional squall. Where the incoming tides nudge the layers of rotting winter ice against the shore, juvenile green crabs hunkered down in their *Spartina* caves have not stirred. But they will.

# CHAPTER 12

# The World According to *Ascophyllum*

I t's the tail end of August and I'm driving northeast to meet Robin Hadlock Seeley, who's going to walk me out along a cove she's been monitoring to show me what she says are the results of overharvesting *Ascophyllum* with a machine. These past few years of attending the Maine Department of Marine Resources meetings to establish a state rockweed policy, and then attending the ones to set aside areas of no harvest, have showcased for many of us who came to listen as members of the public how bunkered down opposing sides to an issue can get, especially when it's business interests versus wildlife habitat.

The Hadlock family has lived on the Maine coast for generations. With long gray-blond hair tied back in a quick ponytail, and an unadorned, oval face, Robin is a portrait of what people might think of as a traditional New Englander, and she has what I imagine is her ancestors' focused

perseverance. As a scientist and an advocate for habitat protection, she came to the meetings, sat alongside members of
the public, and spoke up often, sometimes facing a blast of
pushback. She hardly flinched.

Robin received her PhD in biology from Yale, studying
a species of snail endemic to this coast, and since 2008 she
has worked to limit—more than a few in the seaweed business would say to curtail—the harvest of *Ascophyllum* in the
state. Her argument is that this species of seaweed creates
an inshore forest, an old-growth biome that has evolved
with over one hundred wild creatures that depend on it. To
lose its integrity and quality, she argues, would be to lose an
important element of ecosystem health in a body of water
that has gone through many losses already.

She and her husband, Tom Seeley, built a home in
Pembroke, on Cobscook Bay, a small, tidy cape in a field
by the Downeast shore. She has just wrapped up her work
as academic coordinator for the Shoals Marine Laboratory
on Appledore Island in the Gulf of Maine, which will give
her more time to continue her study of the periwinkles
that were the subject of her PhD thesis and are still one
of her passions, and more time to try to shape this state's
*Ascophyllum* policy.

As I drive, an oversize dump truck pulls out ahead of me,
piled high with harvested *Ascophyllum*. Some strands of the

seaweed trail from the tailgate. Water is spraying off the top of the bed as if the truck were in its own little rain shower. It's an Acadian Seaplants truck heading north into Pennfield, New Brunswick, where the seaweed will be spread onto an old airport runway, turned and dried, and then ground into powders and flakes of different sizes and sent to industrial enterprises all over the world. As the seaweed passes from process to process, many people make money. By the time it's folded into a finished product, it has become a little piece of inshore gold.

Through the openings between the trees, I can see the coves and mudflats of this famous bay. On a map, Cobscook looks as if it were a splash of liquid dropped from a height. It spatters everywhere, crazily irregular, full of little pockets and long points. It is, technically, a boreal estuary with extreme tides, about forty square miles, much of it still quite wild. Deeply muddy, full of rock ledge and salt marsh, and often hidden beneath rolling fog banks, it is a somewhat contained body of water, set back from the open Gulf of Maine by the many convolutions of the land, much the way the Gulf itself is set back from the open Atlantic: Russian dolls, one within the other. According to biologists who have studied it, this bay is exceptionally rich in species, from bacteria and worms to minke whales.

*Cobscook* is a corruption of the Passamaquoddy word *kapskuk*, which, according to Wayne Newell, who has spent most of his life working to preserve and teach his native language and culture, means "at the waterfalls." Its tides stir up whirlpools and swift foaming walls of incoming water where the Passamaquoddies once came to find fish, porpoises, clams, and eels and to pick sweetgrass in the marshes.

I reach the cove, get out of the car, and join Robin for a walk through an overgrown field, down to one of those typical Cobscook shorelines: narrow, some salt marsh on one side, the mud slick and profuse, the bedrock rising out of the mud like a pod of whales festooned with *Ascophyllum* and bladder wrack. There's a small island ahead of us, accessible at this low tide, and to the right of it is the ledge where she is taking us.

Robin has on sensible boots with deep-grooved soles. I have arrived wearing rain boots—an insane choice—and am slipping and falling behind her as she marches ahead. I know she hears me splatting down in the mud, although thankfully she doesn't turn around to check. I am getting to see a lot of *Ascophyllum* up close, and what astonishes me, as I struggle to my feet, is its length. Then it comes to me: of course it's long (some blades seem to be more than two yards from tip to holdfast), because the high tides require it to reach up into them, and a tide can rise in Cobscook as much as twenty-two feet. Reach they do, these monster blades of *Ascophyllum*.

Robin is talking as she strides ahead about the three-dimensional architecture of *Ascophyllum*. This is important, because cutting alters the way it grows back, and it does grow back if its holdfast is in place and some of its initial branching is left. Depending on the nutrient level in the water, exposure to currents, temperature, and other factors, some of which we don't know, it can fill in rather fast. But its original shape, how it reaches up and out to the sides, and how it folds down in layers during a low tide when it is exposed to the air, keeping the under layers wet and briny, is as vital to the community of wildlife around it and on it as the plain fact that it is there, Robin says. It will grow back bushy after losing its apex tip, the one that took it up through the water. In time, one of the side shoots may take the place of that tip, as does a branch of a young white pine when the pine weevil has destroyed its terminal shoot. The branch "volunteers" to be the trunk, turns upward, and leads the tree into the sky.

Robin stops at the ledge. I catch up and stop, too, and she begins to explain to me that we are looking at a careless, destructive, illegal cut: entire seaweeds and their holdfasts have been ripped off, and the ledge has a plucked look.

What resettles the bare rock face, she says, is not *Ascophyllum* at first but the green seaweed *Ulva lactuca*; then the *Fucus* takes over, as it is doing now. It will be some time before the *Ascophyllum* regains its former footing here. Anyone who knows the woods understands this concept of succession. In my woods, for instance, it's raspberries first, then poplar and fir, and finally back to white pine and red spruce. It takes time.

On the way back through the field, Robin tells me, "I've come around to thinking maybe more scientific studies aren't really going to help us, because it's a political thing. The industry looks at their study and they see biomass coming back and they're quite happy with the fact that they haven't, supposedly, altered it. I'm looking at it and I'm thinking I'm not so concerned about biomass, I'm concerned about habitat architecture. How has the height of the *Ascophyllum* changed? How has the branching changed?

"They're businesspeople and what they care about is biomass—the amount that grows back, rather than the way it grows back. They care about what they care about. I care about what I care about."

ROBIN'S STORY OF her understanding of these natural systems begins, oddly enough, in the early nineteenth century. Thomas Say, a famed Philadelphia naturalist who explored this continent when it still was quite wild, visited the Gulf of Maine and studied the inshore snails along it. He described a lovely, small species, which he named *Littorina palliata*. It was thin shelled. Some individuals were bright yellow or pale greenish or dark mahogany brown, and they had high spires like tiny stone-built chapels. His collection of specimens resides today in the Academy of Natural Sciences of Philadelphia, which he helped to found.

"They are beautiful! Absolutely beautiful!" Robin tells me. But when she began her graduate thesis, Say's periwinkle, *Littorina palliata*, which had been the common New England form of the periwinkle *Littorina obtusata* found in Europe, was nowhere to be seen. Remnants of the early populations could be located in drawers in natural history museums, but the living populations had disappeared.

The smooth periwinkle, *Littorina obtusata*, looks very different from the old *Littorina palliata*, although it is also yellow or pale greenish or mahogany brown, but with a thick shell and no spire.

At the Shoals Marine Laboratory on Appledore Island in 1981, Robin, who had come there to work on her snail studies, was walking around the remains of an old hotel, when her eye caught on the mortar of the foundation, which had been built in the mid-nineteenth century. Like any good scientist, she is incurably curious, incurably detailed, and when she looked closer she saw that the mortar was composed of beach rubble, so she started to chip away at it. Before her eyes emerged the high spire of *Littorina palliata*, the periwinkle she thought was no more.

"Twenty feet away in the water," she tells me, "was living the snail I'd been studying, *obtusata*. Right in front of me was a shell of the old *palliata* form. They looked nothing alike! But I knew then that they must be variants of the same species, and I discovered that in a space of a hundred and ten years this snail had made a complete transformation. It went from being high spired and thin shelled to low spired and thick shelled. The question was why."

And that, she says, became her study. When she turned to researching the green crab's journey up this coast, she found that scientists had recorded its arrival near Appledore in 1905. To access the flesh of the snail, the crabs were entirely capable of snipping off the delicate spires and crushing the shells of the young periwinkles.

"If you think of an evolutionary selection force, you've got this massive one entering the environment exerting incredible directional selection. The *palliata* gradually lost their spires and became a lot thicker. They were likely adapting to the green crab."

It was a relatively quick defense, and it seems to have worked. And yet, because she knew that the crab moved slowly up the coast, Robin began to search for remnant *Littorina palliata* populations farther northeast, in Washington County, that had not yet been subjected to crab pressure. She found them in Cobscook Bay, small but thriving pockets of a handful of individuals living in the *Ascophyllum*, and they were just as Say had described them.

"These snails—both the *obtusata* and the *palliata*—are on the *Ascophyllum* all the time. It's their primary habitat and they're wedded to it. They lay their eggs on it, tiny kidney-shaped masses of white dots that turn yellow and then brown and then become the juvenile snails. And they eat it. They eat the seaweed itself and the sloughing, disintegrating parts of it. You can see, if you look carefully, patches on the seaweed that are light colored because the snails have grazed the dark layer of cells off. Very few sea creatures actually eat *Ascophyllum*, but these do."

In 2009, Robin and a number of people who live and work around and on Cobscook Bay organized the Rockweed Coalition, a group dedicated to protecting what they believe is the essential wild habitat that *Ascophyllum* provides. Their issue was not with the harvesters of species of seaweeds for food, but with the industrial harvest of *Ascophyllum*. They were able to convince state legislators that a conservation plan was needed for Cobscook Bay, based on its habitat richness and the dogged persistence of its citizens. They also created a no-cut Rockweed Registry with 568 shorelands off limits to harvest, which simply meant that the owners didn't want their shores harvested. Although no law prohibits a harvester from cutting on these shores, the registry attempts to set up a gauntlet.

"The fact is, when you have a harvest that is taking from what is foundational to an ecosystem, you have to be really, really careful," Robin says. "The DMR doesn't have the resources to manage the harvest carefully. They just don't have the resources to do it right. This is why I have an argument with the people who take the rockweed in industrial quantities. In Washington County a lot of it goes to Canada; in the rest of the state it is processed here. And I have an issue with all of it.

"I believe that if landowners were given the right to own the rockweed at the shore, we'd have a lot more habitat protection than if the state owns it. The conservation groups and land trusts, and there are lots of them—state, federal, private, public—could then legally protect the rockweed if they wanted to. And most of them want to."

WHO OWNS THE rockweed? Is it the property of all the citizens of the state, the way we own the fish in the bays and the ducks in the coves, or does it belong to those who hold title to the land above the tide line?

In the Massachusetts Bay Colony in 1641, a law gave ownership down to the mean high-water mark—a fluctuating tidal boundary dependent on the phases of the moon and the shape of the land—to landowners whose property abutted salt water. But when the tide ebbed, ships had to wait at anchor for sufficient depth to approach the shore. To solve this problem between the ships and the tides, the colonists requested that long wharves and docks be built at the colony's expense out from the land, across the intertidal zone, which was owned by the Crown, and into deeper water so that they might do business at all tides, loading and off-loading wares and passengers into dories and skiffs that they could then row to and from the docks.

Few roads followed the coastline or penetrated the interior. The way north, or down east, into the territory of Maine, was by ship. In this manner trade sprang up, and fish and timber and furs and salt become the pulse of commerce. The colony had no money to build wharves and docks. Instead, in 1647 it drew up an ordinance granting those whose land met the shore the right to construct their own accommodation down to the mean low-water mark, up to one hundred rods in length, each rod 16.6 feet. This meant that the intertidal zone now belonged, in effect, to the upland owner, not the Crown, and he could extend a structure 1,660 feet into a bay if he so wished.

There existed, however, an essential caveat to this new ordinance: the public trust doctrine protects the rights of citizens to navigate, fish, and hunt birds in intertidal waters. Here, in part, is how it reads: "Everie Inhabitant who is a hous-holder shall have free fishing and fowling, in any great Ponds, Bayes, Coves and Rivers, so far as the Sea ebs and flows. . . . Where the Sea ebs and flows, the Proprietor of the land adjoyning, shall have proprietie to the low water mark . . . provided that such Proprietor shall not by this libertie . . . hinder the passage of boats or other vessels." In effect, the landowner owns the intertidal zone, but any citizen is allowed to fish and hunt birds and navigate within it.

In 1820, when Maine separated from Massachusetts to become a state, it brought with it many of the early colonial laws. This was one of them.

Why does it matter today?

Those who want to protect shoreland from an increasing *Ascophyllum* harvest argue that the law gives the public the right to catch fish, which is not the same as a right to cut seaweed, and that fish and birds and boats move through the intertidal zone, whereas seaweed anchors itself to the rocks, becoming part of the upland owner's property through its allegiance to place. Without an owner's permission, any cutting is illegal.

Those whose livelihoods depend on the cutting and selling of *Ascophyllum* argue that seaweed harvesting is indeed a fishery, and that the practice goes back to the beginning of colonial times, when it was gathered for kitchen garden compost and called "seaweed manure." The law of 1647, they argue, protected the public's right to harvest seaweed then and protects that right today.

The Maine Department of Marine Resources has constructed a Rockweed Harvest Plan with hundreds of hours of time donated by the people who volunteered to do the work of creating it, and then at least fifty more hours of meetings for a group to write up a Conservation Plan that would prohibit cutting in some areas that are critical for wildlife. All this occurred before anyone knew whether cutting rockweed was a legal public right, and the plans would go into effect only if no one challenged them.

Most of the shore today would probably be unrecognizable to the lawmakers of 1647, but if present-day landowners believe that the intent of that law supports their ownership of the seaweed, it is only a matter of time before one among them challenges it in court.

Some of the towns around Cobscook Bay are steeped in a way of living, or perhaps it's a quality of character, that reminds me of the best of the Maine past. When I turned off the Old County Road, I parked in the driveway of an old farmhouse that serves as the town of Pembroke's public library.

I was here to sit down with Ken Ross to talk about *Ascophyllum* cutting, the Rockweed Coalition, and his thoughts about who owns the rockweed. I opened the door to the main room and was met by spirited laughter. A group of knitters was gathered around a large table, their hands busy with needles and yarn. It was a pleasure to meet these women on a bright August afternoon, and it seemed to me that they had known one another for years, which reminded me of the days when small-town coastal life was animated and productive. They were vibrant places by a giving sea.

Ken and I stepped into a quiet corner in the children's section of the library and sat down to talk. He had attended many of the DMR meetings, and I remembered that he always seemed somewhat perplexed during them and kept coming back to one question: Why are we talking about a harvest if we don't even know who owns the rockweed?

He grew up on the shore of Red Beach, a community fifteen miles north of here and directly across from the island where Samuel de Champlain attempted to establish a colony but instead spent a disastrous first winter in the New World.

"My family and I have a lot of shore property, and we've got the highest tides in the world around here, which means we have a lot of seaweed. This is the mother lode we're talking about. I'm in the unusual position of having deep roots here where the rockweed cutting is going on. I know the way local people think because I'm one of them, and I've taught environmental issues for twenty-five years, so I know that environmental issues are almost always political issues. If it's an issue, it's a fight. And if it's a fight, it's politics."

Ken went to graduate school at American University in Washington, DC, studying political science, and taught at Adrian College in Michigan. He's retired, and for half the year he lives in Maine and travels often to this library to work and read and get Internet hookup. When I ask him how the Rockweed Coalition started, he says, "A few people noticed the greatly escalated cutting going on, which began when Acadian Seaplants expanded in this direction, down from Canada," he said. "People hadn't given much thought to it before, but they started to worry about the ecological effects from this large-scale cutting. There were complaints going to the Department of Marine Resources and also to the state legislators because the cutters were coming in along the shore without asking permission, and if you went out and asked them to stop, some did, but some didn't.

"It grew into a storm of protest, and eventually the state legislature wrote a law in 2009 that provided some means of protection for Cobscook Bay. I'm sure that the public furor was a factor in bringing it about. Would the legislature have done anything if no one complained? I doubt it."

He is a great admirer of Robin Hadlock Seeley for her intelligence and her tenacity. "Robin was one of the founders of the coalition," he says. "It consists of a finite number of people. I help all I can and so does my brother, and I know that it's exhausting and expensive and consuming work. It goes on all year long. The issues are complex and the coalition operates on a political level, a scientific level, a public relations level. There's always something happening. How many people are going to do that kind of work on a volunteer basis?

"There's only one person who has come near that. It's Robin. The coalition has some lawyers in it, some scientists, some smart people who help out, but you need someone masterminding it all the time or you get blindsided. You miss opportunities. Robin doesn't let that happen. To do what she does, you have to devote your life to it."

When I ask him why he got involved, he says, "If we're going to exploit a natural resource, we should pay attention to the examples, many of them local, in which we've depleted resources and there's nothing left. Which means no jobs either. It's not just the destruction of organisms and the loss of biodiversity, but it's the jobs. It's both. In this county we're desperate for jobs. But before we do something like this, the sensible thing is to find out what the effects will be, not just go ahead and do it and then find out what the effects *were*.

"Anything that provides jobs is almost immediately seized upon by lots of local people. And when you take an issue like this that features a common plant with the word *weed* in its name, the average person thinks, Who cares about a weed? And when you cut a weed, what does it do? It comes back up again. So the thinking is they can do this cutting just like mowing a lawn.

"You can understand the thinking: We didn't realize we had a resource here, and we can make some jobs out of it. We can take money home and feed the kids. And if no one is complaining about it, everyone's happy. So it got momentum, it was so well accepted that it was difficult to catch up to say, Wait a minute, what are we doing?

"The tendency is to say all these environmentalists are just rich people from outside. But the environmentalists I know want to see sustainable jobs, and in the Rockweed Coalition's view, there hasn't been a proper scientific study done to justify the mass cutting."

And what, I ask, does he think is going to happen? He laughs, a bit ironically. "I don't know what's going to happen, but the ocean, if given a chance, has highly restorative powers."

While writing this book, I have had the opportunity to read scientific papers on the effects of *Ascophyllum* harvesting. They present disparate results, and it's no wonder. The variables are many and hard to measure. A biologist doing a study on this species of seaweed has to contend with the tides, the weather, the time of year, the creatures who move through the seaweed quickly, taking what they need, the ones that slip off the blades when disturbed and roll away, and the ones that hold on tight. She has to factor in the juvenile fish of many species and the green crabs and the hermit crabs lugging their heavy houses, and periwinkles and amphipods and brine shrimp and sea worms, all moving at different speeds through the *Ascophyllum* at different

times of day and night, tide and weather. How can biologists measure all this and come up with an answer that can
be duplicated in a subsequent study? There are hundreds,
maybe thousands, of variables. Which ones do you measure?
Which ones do you count?

AS SEAWEED HARVESTS take the place of lost fisheries in many areas of the world, they present some of the
same issues we have here: the growing desire of coastal people to take good care of what's left, a need for more education and study, and an acknowledgment that the oceans in
our lives are in trouble.

In Ireland, a country that has been cutting seaweed for
hundreds of years, citizens who live by the coast are educating themselves about the issues of seaweed conservation,
as they are doing here in Maine. An organization called
Coastwatch Europe is studying seaweed protections and
uses in the country and has initiated a curriculum to teach
students of all ages what seaweed habitats provide and why
they are basic to coastal health. In the classroom, and then
out along the shore, students join scientists who show them
how to identify species (Ireland has more than 500 species of
seaweed that grow in its inshore waters, compared to the 250
species in the Gulf of Maine), how to recognize the seasonal

reproductive periods of different species, how to assess high-value sites that need special protections, and more.

The organization is building a constituency of citizen scientists who understand both the environmental values and the harvest benefits of seaweeds along their home coast. With seaweed businesses, some of them international in scope, making substantial inroads into the Republic of Ireland, the years of traditional small-scale harvests are almost over, and it's past time, both there and here in the Gulf of Maine, for policies that everyone can agree on.

Perhaps we can agree on this: we can start to learn from the past. A person I admire for his work for change is Dr. Gavino C. Trono Jr., who has spent his life studying the seaweeds of his home country, the Philippines, with the purpose of connecting the abundance and diversity of edible native seaweeds to the impoverished people of the Philippine coasts.

Dr. Trono is now eighty-six years old. His work has spanned over fifty years. The results are far from perfect, it's true, but they are a beginning, better than before his nation took note of his studies and his ideas about how to apply them. The deep poverty of the villages along the shores was principally the result of overfishing and horrific fishing practices, including the use of dynamite and sodium cyanide to stun and catch fish that live among the coral reefs. These practices lead to shattered, dead corals and few fish to eat or sell, and human hunger and ruined bays took the place of what was once one of the most abundant areas of sea life anywhere. When Dr. Carl Safina visited those same coasts in the 1990s, he wrote that he expected to see Paradise but found instead a Paradise Lost.

Dr. Trono has introduced seaweed aquaculture to the ruined bays, providing coastal families with work that brings in both food and income. There is much more to be done. Monoculture farms are vulnerable to overcrowding and stagnation. When the crops are packed too tightly for the tides and ocean currents to infuse the bays with clean water, as Tollef Olson has explained, the result can be anemic seaweeds vulnerable to disease that spreads throughout an entire crop. Nor should seaweed farms impede the lives of native species that use the bays, and this, too, is something that requires work in the Philippines.

Even now, Trono continues his efforts to teach the value of functional bays and healthy crops within the 154 square miles of Philippine seaweed farms, and in 2014 he was awarded the title of National Scientist of the Philippines for his life-changing gifts to his people.

In 2016 the UN issued a report of concern highlighting the unregulated seaweed farming in the Philippines, South

Korea, Indonesia, and China. The report noted that in this sprawling, fast-growing business, seaweeds of a number of species are being moved out of their native waters to new farming sites, and some arrive carrying pathogens, wiping out whole aquaculture enterprises.

These farms can save people's lives, as Dr. Trono has demonstrated, but only if we learn from the past, and that means changing how we interact with wild environments.

ON MY WAY home from Cobscook, I stopped in to talk to two people who have been involved with rockweed issues ever since Acadian Seaplants brought their boats and rakes into the bay and hired subcontractors to cut the *Ascophyllum*.

In late morning I walked the driveway to the farmhouse in Whiting where Julie Keene lives with her partner, Adam Boutin. Their fishing boat was up on wooden jack stands. A field lay beyond it, and around the outbuildings by the house was the flotsam of a working life both on the water and in the farmyard.

Julie and I sat down at her kitchen table. Hers is a fishing family going back generations, hardworking, outspoken people. Her worry is that the rockweed harvest may diminish other fisheries. She used to harvest periwinkles, taking them in the cold season, when they hunker down together in rock crevices beneath protective layers of *Ascophyllum*. She'd heave the seaweed to one side and pick the "wrinkles."

She tells me she was one of the first people to join Robin as a member of the coalition.

"The people that stay here, so many of them generational, rely on that ocean for wrinkling, for scallops, for clams, for lobsters, for whelks—everything!

"Then we started looking at who Acadian Seaplants was. They came over here and the State of Maine gives them non-resident harvesting permits. But I can't go across to Canada and harvest one periwinkle, one clam. Nothing! There's immense frustration. Immense sadness. Years of fighting this, of trying to make this rockweed harvest go away.

"When the sardine factories closed here," she says, "Adam and I were buying periwinkles and clams in our truck as middle people, for a business in Jonesport. Because the factories had closed, I had people coming to us that had never picked periwinkles in their life but the work was gone for them, and out of desperation they came to us. It was the saddest thing to see us lose our identity with the factories gone. So many people out of work. And then infectious salmon anemia hit the salmon pens.

"Everywhere I went, people asked, What are we going to do? What are we going to do? I was buying their periwinkles.

People coming with duct tape on their boots, mufflers wired up on their cars. It made me mad."

She told me that for years she and her family harvested periwinkles, sorting them on a mesh screen so that the small ones would fall through into buckets. They kept the bigger ones to sell. Then Julie would take the small ones over to a cove with lots of seaweed growing in it, and dump the little ones back in.

"It was like farming," she said. "After a while, we could go in and harvest them again." But seaweed cutters came into the cove, and now, Julie said, there are only a few periwinkles left.

"When I go down to the beach I can feel the lives of my ancestors, and I'm scared to death with what's happening."

CARL MERRILL IS the former director of Suffolk University's Friedman Field Station in the town of Edmunds on the shore of Cobscook Bay, where he taught college and high school students about marine life from May to October. The field station recently closed, but he remembers coming as a student in 1976 when he was eighteen years old, because this is the place where he fell in love with ocean science and never looked back.

"There was such diversity, it was spectacular!" he said when we sat down in his cramped office filled with papers and books. He told me that diving into the cove below the field station was a life-changing experience for him back then. "Today, students encounter a different environment. They're impressed, but it's dramatically different. A lot of species that were here aren't. And those that are here are fewer, often smaller."

When I asked him why he thinks this has happened, he said, "Fisheries pressures changed, and technology developed. The good thing about fishing is it provides a livelihood for a family. It makes sense to be harvesting the resources of the sea. But the way we've been harvesting doesn't make sense. We need to fish on a smaller scale that benefits the person out on the water, not so much the people further up in the industry. I love the idea of fishing. But we've pushed it too far. So much horrible waste goes on out there, and when we overharvest we hurt the fishermen as well as the fish.

"And I am very reluctant to see large-scale harvest of the rockweeds," he continued. "They provide food and shelter for so many organisms. We need our finfish species—the cod, haddock, flounder—to recover enough for us to find out how their young use the rockweeds. That means we have to be patient. It doesn't make sense to harvest this resource if the harvest impedes the recovery of a resource that is far more valuable economically to this community.

"But the people who are involved in managing ocean resources, I have to take my hat off to them. It's very complicated to manage these resources. It's an inexact science, and they have to deal with all the socioeconomic situations, too."

I HAVE SAT in on a number of events where Robin Hadlock Seeley presents her educational program on *Ascophyllum*. In each case, she has a polished PowerPoint, with gorgeous photographs of seaweeds and fish and ducks and shorebirds, and I have seen how closely her audience attends. People who love the places where they live often don't know much about the wildness in them, but they want to learn. Robin leads them through the stories of interconnected lives.

I've seen her start with rockweed as habitat: where it grows attached to the rocks, where it floats loose out at sea, and where it washes up, tumbled, against the shore. For each manifestation she points to how it's used by a variety of wild creatures, from juvenile cod to purple sandpipers, and in doing this work she attaches the world of *Ascophyllum* to the world of the people in front of her.

She ends with this: "Should we be harvesting habitat? Is this really what we should be doing?"

On December 11, 2015, Ken Ross, along with his brother, Carl, and a group of homeowners, filed a complaint against Acadian Seaplants Limited in Washington County Superior Court. The purpose of the suit was to determine who owns the rockweed in the state of Maine, and on March 16, 2017, the court declared that the rockweed belonged to the shorefront property owners and is not a resource in the public trust. Acadian Seaplants has appealed, and the case will proceed to the Maine Supreme Court.

When I called Ken to ask what he thought, his response was measured: "Determining who owns the rockweed is necessary," he told me, "but for the long run, we need science-based limits on the taking so the interests of the homeowners, the public, harvesters, people who fish, and the wildlife will not be diminished."

Until these rights are resolved, both the new Department of Marine Resources Harvest Plan and the Conservation Plan for *Ascophyllum* are on hold.

The harvesters of edible seaweeds are not a part of this suit. They continue their cutting and drying and packaging and selling in ways that are reminiscent of the earliest fishing culture on this coast: boat and fisherman, nothing fancy, nothing that takes away from the harvester's direct relationship with the water.

As THE PUBLIC discovers how basic and essential *Ascophyllum* and the *Fucus* species of seaweeds are to the lives of the bays where they grow, phycologists are studying the effects climate change may inflict upon them.

With the ongoing threat of a warming climate, what will happen to *Ascophyllum*, along with the two primary species of *Fucus* that with it create the canopy-forming rockweed forests? Recent studies suggest the possibility of extinctions of these seaweeds at the southern edge of their range because of warming, and a thinning out of genetic diversity, as well as a slow spread northward as Arctic waters open up for them. But at what point the rockweeds may reach the limit of their ability to endure the prolonged dark of an Arctic winter is unknown. What is more certain is that if they go, they will leave behind exposed shorelines, with little to deflect the battering of storms.

Scientists cannot predict whether the warming trend will speed up, continue at a consistent pace, or perhaps even diminish over the coming century. Nor can they be sure whether the snails, scuds, fish, and other lives that depend on this particular group of seaweeds might be able to adjust to future changes. Will some of them be forced to adapt to life without these foundational species? If so, which ones will they be? And will those that cannot be lost? Or will

# CHAPTER 13

# Tending the Wild

I'm sitting in the bow of a sixteen-foot skiff heading out to a small island near Isle au Haut, seven miles offshore of Stonington, the old fishing and quarrying town at the very end of Deer Isle. At the stern, Micah Woodcock is piloting us steadily across open water. Behind us, tied firmly to a painter, bobs a nineteen-foot dory. The early morning light glints off the water, softened by a thin veil of retreating fog.

We are going to pick up the kelps he has been drying on an island over the summer, and he's promised to take me around the ledges so that I might see where he harvests them. Micah, twenty-eight years old, tall, and rangy, has eyes so deeply blue green that one could be forgiven for thinking that they look a lot like pools of seawater.

He cuts kelps—*Alaria*, horsetail kelp, and sugar kelp— dries them out on the island, either on structures that look like industrial-strength clotheslines set up under the sun, or in a barn with dryers operated by a generator. He'll put some of the kelp blades through a food mill, selling the flakes, and what he calls "whole leaf," through his website. In simple brown packages printed with the name of his business, Atlantic Holdfast Seaweed Company, he ships his products out to buyers. Occasionally he sells the long blades to other processors after he dries them. At the midtides, when the kelps are too far underwater, he picks dulse and laver and Irish moss. It is physically demanding work, and it requires patience, precision, and a deep knowledge of the seaweeds along this coast and where they grow best.

We are passing the quarry islands, some with rusted cranes still standing where, generations ago, they once lifted blocks of fine-grained granite onto barges. We make our way past islands with summerhouses, and some with nothing on them but spruce and a few pines, gulls casually strolling the ledges, an occasional bald eagle standing at attention on the branch of a tree.

The hundred-acre island we come to has a cove where the wind drops off and the water lies still. Before boats were motorized, people made homes on islands like this one and raised families. Living on islands brought them closer to the fishing grounds, but the families were isolated, and when they were able to move quickly around the bays in motorized boats, most of them gradually abandoned island life. The mainland had more people, more stores, schools, and churches: more small-town community. A fishing family owns this island and has offered Micah the use of it.

Micah ties up both boats. We climb a ladder onto a rickety dock. Before us lies a field with a cellar hole where a farmhouse once stood. Four old sugar maples surround it. At the island's height sits the boathouse he has turned into a seaweed-drying shop, and beyond that, within some spruces, I can see the cabin he stays in for about half the year. We head up to the boathouse, where he sorts out the large plastic sacks of kelp he's cut and dried, with dates and the species of seaweed written on each sack, and we set down our gear in the cabin and then head back to the dock. He's going to show me where he cuts kelps.

It took him two years to select good harvesting beds before he began to cut, and during that time on the water, he was learning about how this bay works, where the ledges are, how the currents run, and where the kelps flourish.

"I harvest during the new moon and full moon tides and generally spend the weeks in between processing what I've harvested," he tells me. "I start in late March and early April and finish up sometime in September or November."

When he reaches a ledge in his skiff, he tells me, he sets the anchor and rows the dory he's towed, pulling it up out of the tide onto the ledge, then slips into the water in his wet suit with his cutting knife.

"Getting to the ledge can be difficult, and most of them don't have a protected part. On the really exposed ones I'll bring bushel baskets, which I've lined with nets. I'll fill the nets with kelps, and if I have no safe way of getting the baskets into the dory, I'll float them off with the tide and chase them down with the skiff afterward." Micah often works alone, and the description of the job, which he offers as if he were describing a desk job, only underscores its dangers. In order to make this career of his succeed—flourish, really— he needs to go where the best kelps grow, and those places he's chosen, I see now, are wave tossed, with quick currents, and the ledges are jagged and slippery.

Micah was born in a little village in the northwest corner of Maine. When he was two years old, his parents, who were missionaries, moved their family to Greece. For twelve years, Micah grew up learning all about boats and the water in a small Greek harbor town. The family returned to this country and he finished high school, but he had no interest in going to college. Instead he went into the world to learn what he wanted to know more about. He was looking for people to teach him, as the old fishermen in the harbor in Greece had done for him when he was a child. This coastal life, where the physical work is hard and the knowledge and craft of seaweed harvesting takes time to learn, still provides a place where people like Micah can fit into the world with dignity and intelligence, using all they've learned and studied, well educated without a college degree.

Eventually, Micah met Larch Hanson, one of the first people on this coast to harvest and sell edible wild seaweeds. Larch began his business, Maine Seaweed, at about the same time that Shep and Linnette Erhart began theirs, over forty years ago. In his midseventies now, tall and rod-thin, Larch is still taking his handmade boats out along the coves near his home to harvest in season.

Micah says, "He's one of the fathers of the modern seaweed industry here in Maine, which is still quite small but getting a lot more attention." For years, Larch trained young people to do this cutting along with him, instilling in them the benefits of careful wild husbandry. He taught them that they can make a living and feed themselves well without harming what's left of the natural world. It has been his life's work.

Micah was part of Larch's regular crew for about a year.

"He's a good teacher," Micah says. "I asked him, 'Is there room for other people to be doing this kind of harvesting?' And he said yeah, there are other places here where you could set up a business like this.

"I learned from him what I needed to know to start a business and to tend a wild garden. That was what attracted me to the work: tending and harvesting a wild resource. Larch has been in the same place doing this harvesting for a long time, and he actually has, as he likes to say, the memory of place that you get from being somewhere for so long.

"*Stewardship*. I like that word," Micah says. "It means taking care of something that takes care of you. But there aren't a lot of places left on the coast to set up a business like this. For instance, we just came through miles of water to get here, and there's no edible seaweed to harvest in those miles. How much of the Maine coast has harvestable sea vegetables growing on it? The number's small.

"I don't think that a poor tending of the commons is a good reason to privatize it. I'm not opposed to rock-weed harvesting, for instance, but I don't like the way it's being done in some places, and I think there are places where it shouldn't be cut at all. I want to see people harvest rockweed who actually live here and who harvest in a responsible way."

Micah has twenty-four separate harvesting sites scattered around a large area, and he knows them with the sort of specificity that he admires in Larch Hanson. We reach his first ledge about forty yards to the west of a small island. Within all this water surrounding us, it suddenly appears straight up, black rocks protruding like giant broken teeth a few feet above the water. The surf is hitting the rocks from all sides, sending up a froth of whitecaps as we make a broad circle around them.

BACK AT THE island, as the tide comes in, filling the cove and raising Micah's boats, we carry down the sacks of dried seaweeds, stacking them carefully on the dock, and then, when they're all down and stacked, Micah sets them in the empty dory, placing them so that they do not cramp one another. Then he covers them with tarps and ties the tarps down, and we begin the seven-mile trip back to the mainland, again passing the islands and the history and the overlapping visions of a home place.

A WEEK LATER, I have come to an evening sea-weed-cooking class at the Island Community Center in Stonington. The weather is warm, the light drowsy with the end of summer. An old Quonset hut contains the kitchen and tables, and Micah will be teaching the class as a part of the Edible Island series. I notice on a poster taped to the front door that next week's class is how to butcher a pig, with the pig in attendance, and I'm grateful I haven't gotten my dates mixed.

About twenty people have signed up: a businesswoman with a home in Stonington, a young woman who works on a farm farther south, two doctors, a professor emeritus from the School of Earth and Climate Sciences at the University of Maine, his wife, and others, a good-humored crowd, ready to try something new. Not only will Micah teach us how to cook seaweed tonight, but he's prepped to give us a quick overview on pretty much everything and anything about seaweed that we'd like to know.

He starts with a distinction that applies around the world: there are two kinds of seaweed harvesting. One is industrial, in which farmed and wild seaweeds are cut and processed and sold to companies that make all sorts of products, including additives for human food. The other is edible seaweed, or, as some say, sea vegetables, cut wild or grown in aquaculture sites and minimally processed.

He has brought along jars of dried edible seaweeds he's cut, and lines them up on the counter in front of him. He tells us both the scientific and the common names for each. But, he says with a laugh, "the phycologists have been changing the scientific names faster than I can keep up."

He passes around samples of dried laver and dulse and Irish moss to try. Most reject the Irish moss, calling it "fishy." Then he takes some dulse and laver and spreads them on a pan, covers them with a splash of oil, and sticks them in the oven. They crisp and taste a bit like chips.

"I call them the gateway seaweed. Almost everyone likes them," he tells us.

"The bladder wrack I've brought is used medicinally for the thyroid. It can be ground into a powder and put into capsules—it has a bitter taste. It's good for burns, too, good for your skin. It's soothing. You can put a handful in a net bag and drop it into your bath and the hot water will pull the gel out.

"Irish moss soothes the pain of shingles. You cook it, get the gel out, and use that as a poultice.

"Some seaweed varieties wash up along the shore. If it washes up, it's generally not good to eat. The smell and taste are off," he says. And yet, Irish moss, washed up and bleached white by the sun, is an exception.

"Going into the fall harvest, which starts soon," he tells us, "the sugar levels go up in the *digitata*. After I dry it, if I introduce just a little bit of humidity, it tastes profoundly sweet. Delicious!"

Tonight's menu is sugar kelp noodles with garlic and butter and mussels, and an *Alaria* salad with blueberries, cucumbers, maple syrup, olive oil, and apple cider vinegar. He drops blades of sugar kelp into a big pot of water to cook, *Alaria* blades in another big pot, and sets them on the stove.

"Sometimes with *Alaria* I pour boiling water over it and let it sit through the night. You want to save that water and the cooking water because that's where some of the trace minerals and the iodine are. You can use them in a soup base. The big thing about seaweeds you won't get with land-based plants is iodine, and a lot of people in the United States are deficient in it.

"One of the most important reasons to incorporate seaweed into your diet is trace minerals. Seaweeds concentrate the naturally occurring trace minerals in ocean water. It is estimated that the human body needs fifty-five to sixty trace minerals for optimal functioning. Seaweeds have them all."

The *Alaria* and sugar kelps are ready to prepare. Micah rolls the sugar kelp up and cuts it into chunks, and we set to slicing them into thin noodles and chopping the *Alaria* as he oversees us, answering questions.

"I eat a lot more seaweed in the winter, in salad and as noodles," he tells us. "It stores frozen or dry. You rehydrate it and you have this delicious green vegetable. I add the kelp mix I make to rice and beans and soups, using it sparingly. You could feed people foods with this kelp mix all day long and they'd have no idea they're eating seaweed, but it makes things taste a little bit richer, fuller."

He sautés the noodles with butter and garlic in a large pan on the top of the stove. Soon it's time to eat. We gather around the tables, serving ourselves the thinly sliced noodles—a gray-green color, like spinach noodles—with a cluster of bright orange mussels on top, and we pass around and take some of the shimmering salad. The *Alaria* is a radiant green, and combined with the marine blue of the fresh blueberries, it is almost too beautiful to eat.

THE CURRENT IN the bay streams past the fourteen-foot boat I'm in like an undulating velvet carpet: smooth and fast, almost silent except for the sound of the water brushing against the hull, and the small sounds the water makes when it encounters impediments such as the cobble along the island shore, not more than five yards away from us.

Andrea DeFrancesco tosses in the anchor with a loud splash. She and I are harvesting kelps. The tide is falling, but in this narrow channel between island and mainland the water is moving both ways. At its center, it is still going into the bay behind us, as if it hadn't gotten the low-tide message yet, but along the sides, along the banks, it's leaving.

Andrea grew up here and went away to Hampshire College for three semesters, but "college was not for me," she says. Instead she worked on boats out on the West Coast and then came home again. Now in her midforties, she is an advocate for social justice and a healthy environment and has worked over the years as a carpenter and landscaper but found her real place harvesting edible seaweeds. She worked for about seven years harvesting for Shep Erhart, and then for someone who eventually sold her his seaweed business.

Last winter I sat and talked with her as she sorted and cleaned the dulse she had picked during the summer and dried on racks in her drying shed. She sells to individuals, other businesses, and stores, but "I'm not a businessperson, really," she said. "I'm a harvester with a business." It's named Ironbound Island Seaweed, and she's owned it for two years now.

She harvests in this channel, and also keeps a larger boat at Bunkers Harbor, just below her house, in South Gouldsboro, and takes it out along islands: Stave and Jordan, Porcupine and Cod Ledges, Crow and Ironbound, and Egg Rock.

Before we climbed into her boat to harvest, we sat at a picnic table and talked, and it wasn't long before a neighbor of hers, a newly widowed retired fisherman, walked down, took a seat beside us, and began to tell me how he watches out whenever he sees Andrea in the channel. Sometimes he'll lose sight of her, he said, and he'll come down to the shore to make sure she's OK. After we talked for a while and he left, Andrea told me, "I think it's a good time now to look at the story of our fisheries that we said we'd never destroy and then did. We can put some kind of framework in place to keep that from happening to the seaweed harvest.

"I have learned to regulate my cutting. It's having a sense of it, not just wandering in blindly. I know about growing things and I've had a lot of input from people who know how to harvest seaweed. You're not supposed to take more than a certain percent of a bed, but mostly I regulate myself by thinking of what else is using the seaweed and what impact I'll have on them as well as the impact I'll have on the ability of the seaweed to reproduce itself.

"I want to leave swaths that are completely intact so they provide protection for other lives. If I just went in and thinned everything, like tree thinning, the beds would be too sparse for them.

"One of the best things about seaweed harvesting is it's not tourist dependent. I'm harvesting something sustainably that's here. I am providing it to people because it's important to their health. It's something real, and people are grateful for it."

She told me that the harvesters who cut edible seaweeds have so far managed among themselves to allocate different bays to different people. She works the bays between Larch Hanson's harvesting and Micah's.

When she goes out into the larger bay, rather than here within the channel, she pilots a sixteen-foot scow and trails a small punt. Often she goes out with a helper, but just as often she goes alone. Sometimes it's dark, and she picks her way between lobster buoys. It's almost always cold in the water out that far, and she wears a thick wet suit called a Farmer John, and over it she pulls on a step-in jacket. Today she's in shorts and a T-shirt, but before we pushed her boat out into the tide, she put on a summer wet suit, and wet suit booties, and both of us put on gloves. We piled plastic baskets in the boat, pushed it into the water, and got in. She started the motor, and we were off.

Like Micah's, her season starts in early spring—in March. She cuts in the negative low tides, as he does, which means around the full and new moons.

"Sometimes if I'd be harvesting at sundown, I'd go home and hang up my seaweed and then pull off my wet suit. The next morning the suit was as cold as ice, and I'd have to pull it on and go out again. And sometimes when I start the harvest for the year, there's still snow on the ground. I'm yelling to myself, What in the hell am I doing here? I have to shovel snow to get my boat out."

About a foot underwater, the swift current is ruffling the sugar kelps' long fronds. They look to me almost like eels, many eels all swimming in the same direction. I lean over the side of the boat, grab a kelp by its stipe, and slash it free with one of the big sharp knives she works with, then quickly lift it out of the water. Because it's September, the kelps have some deterioration at the tips: epiphytes, a few holes from creatures feeding on them, some sloughing material that is a combination of what the seaweed makes itself and the bacteria that live on the fronds. We slice the worn parts away—Andrea patiently showing me how—and toss what's left into the baskets. Then over and over again, another frond, another slicing and lifting, as we work our way along the island shore past an immature herring gull too interested in probing between the rocks to be bothered by us.

When we get back to shore, we carry the baskets up to her truck, and then I follow her truck in my car to Bunkers Harbor because she wants to show me where she launches her larger boat to get into the deeper water. As I drive behind her, I am thinking of one of the things she's told me this afternoon: "This is where I grew up, and the ocean has always been a sustaining factor in my life. I pay attention to it, and I like the tide dictating my movements. What I love about this work is that I'm not doing any harm."

We walk down to the public dock as she explains what it's like to take her boat, moored off the dock, out of this harbor. "Sometimes I'm scared," she says. "I'd silly if I weren't. But what you find out there when the tide drops is beautiful, the quality of light and the sounds, and no one else is seeing and hearing this but you."

# EPILOGUE

*For better or worse, we shape the future in the present.*
CARL SAFINA

In early March I like to walk the rocky headlands or take the mail boat around Isle au Haut to get a last look at purple sandpipers and harlequin ducks before they head north.

Most days it's still bone cold, the surf rough, and I search for these birds that are hard to see among the rocks and seaweeds and churning water. But they're there. They are tough when it comes to surviving where they choose to live, and vulnerable, especially now, with the changing climate. As I walk the edge of the land or stand on the boat, I see them doing what they have done for a long, long time.

Harlequins, named for the male's dramatic pattern of white, black, blue, and rust-red plumage, are small ducks, listed by Maine Department of Inland Fisheries and Wildlife as threatened because their numbers here are few. They swim so close to the *Ascophyllum* and so close together that sometimes they look like one duck with many heads within a line of breaking surf and seaweed. Then, as a wave disassembles into foam, they disappear. Nothing. They're gone.

They pop up yards away.

Once, in rough water, with a wave lifting high and cresting before it crashed onto cobbles, I saw a single male harlequin swimming horizontally through it. The water was a light translucent green, and the duck shot sideways within it like a bullet.

Purple sandpipers are even harder to see. They're the color of the rocks, and if you spot one, it is as if a bit of rock suddenly shifted—and then you see it: a purple sandpiper stretching its wings. That is how they come into focus, a flock just inches above the tide, some feeding in the drapes of winter *Ascophyllum* or across the deep orange mats of *Mastocarpus stellatus*, others hunched down, resting, and some walking in that stolid, matter-of-fact gait of theirs, as if they weren't at the lip of pounding surf.

They are site specific. If you can remember where you saw them last winter, you'll probably see them there again this one. At Schoodic Point on the Gouldsboro Peninsula, I found them on a favorite midtide rock, a few yards offshore, a tiny handful of birds, no more than six. One seemed to have only one leg, but sandpipers notoriously lift one or the other leg into their belly feathers and will often hop around on the other. I waited to see this bird settle down on both legs as its flock mates were doing, stepping among the barnacles, oblivious to the scouring wind. The bird never did put down a second leg, but hopped forward, feeding with the rest.

Suddenly they lifted all together, and flying as one, tilting in a tight flock from side to side, they vanished over Little Moose Island. That, I thought, is what these birds do—they get on with it, one leg or two, within the changing tide and the wind.

The number of purple sandpipers is dropping along this coast that is their winter refuge. People who study them believe it may be because melting permafrost is swamping their nests along the shores of taiga lakes on Baffin Island and around the western side of Hudson Bay, where they breed.

Lindsay Tudor is a biologist with Maine Department of Inland Fisheries and Wildlife. Her job is to oversee shorebirds and some species of ducks. In the winter months, it's purple sandpipers and harlequins, two species whose habitat is rock, water, and seaweeds. I have watched her at the meetings to set aside no-cut areas for *Ascophyllum* to protect these birds, and she is a gentle but firm presence, willing to state her principles as clearly and as often as she thinks she needs to. She spends the winters out in open boats, dressed in layers of gear that make her look as if she were about to blast off to the moon, binoculars at the ready, hour after hour counting the birds she has come to love.

I WROTE THIS book to explore the complexity and beauty of a wild inshore system to which, I believe, we owe a great deal, but the people I have learned from have given me more than that. They startle me with the depth of their care for where they work and the dedication to the work they do. Sometimes it seems to me they have made themselves part of the wild systems they work within, as witness and voice.

What keeps me guardedly hopeful about the future is not only the astonishing resilience I have seen in much of the natural world, but the plain fact that the people I write about give years of their lives to it. Making a choice to be here, to stay, is a choice of consequence.

And they are believers in the value of coastal communities. They understand that you can't protect a wild ecosystem if the people living beside it feel diminished. Coastal people need jobs that require of them the sort of skill that the old fishermen once had, but the jobs won't come unless what's left in our bays and deeper oceans is vigorous and plentiful. Jobs aren't made out of nothing.

Those old fishermen didn't own a lot of fancy equipment. They learned the important details about the world they worked in, not only from their fathers and grandfathers and other fishermen, but from their days out on the water. Then we lost the fish, and the heart all but went out of the inshore waters and their fishermen. But the fish didn't disappear on their own, and we can't erase the fact that dragging still destroys the bottoms of our bays, trashing complicated and life-giving seaweed habitats and rocky beds. Nor can we overlook the harm trawling can do, nor the careless waste of throwing bycatch overboard.

When I first arrived to make my home on this coast, I learned that the fishermen down at the harbor knew things, all sorts of little and big things about the water. Their work was a craft, some might even say an art, and when they went out in their boats they read the water, the sky, the tides. They'd go out, the best of them, with a sense of self-respect that came from this attentive and hard-won expertise, this relationship with a wild place that is both dangerous and giving. But I don't romanticize them, for they endorsed the myth that abundance would last forever, as did so many others, even when all the signs they saw—with their educated eye—pointed to the plain, hard fact that it wouldn't.

What we can learn from the fishermen of the past is this: The wild tests people, hones their skills, offers them a way to be in the world that can give them pride. If we can take that sort of learning and bend it toward scientific inquiry, careful harvest, and resource protection, we will be creating something fine. The purpose of regulating the wild seaweed harvest in the state of Maine—and in the world at large— is twofold: first, to identify and protect essential habitat; and second, to build a new model of how to manage ocean resources that doesn't edge them toward oblivion.

People who spend time at the shore where they live, who see the birds coming back and see them leaving, who make an effort to learn about the complicated life of the water in front of them—they, too, are part of the inshore world. They bond to place. And wild places will teach us, if we let them, if we pay attention.

We may not know everything, and we don't always agree, but we know more about what went wrong with our fisheries than we did before, and we know a lot about sea-weeds, though not enough. If we can quit the old practices that aren't working, rather than tossing the blame elsewhere, we may be on our way to welcoming back some of the wild diversity we lost.

The inshore world along countless coasts in so many countries includes seaweed forests, sea grass savannahs, mangrove edges, and coral reefs. These places have poured out astonishing riches—just about everything that leads to everything else the oceans need. They are the equivalent of marshes and swamps, forests, short and tall grass prairies, taiga and tundra. In E. O. Wilson's book *Half-Earth: Our Planet's Fight for Life*, he writes that the earth's shoreline is about 356,046 miles long, and my guess is that doesn't include all the islands. His point is that these are many miles to squander or to save.

Building policies that honor the complexities of inshore habitats is the way to protect them. One conservation tool, among others, is to set aside large tracts off-limits to exploitation, and then to train community people as wardens, researchers, and wildlife guides, whose job it is to oversee them. Nothing about preserving coastal waters into the future will be cheap or easy, no matter how it's done.

The natural world is shifting, which means we need to hurry to fix what's broken and pay close attention to what's coming next. The wild must come first if we are serious about safeguarding a future to hand over to our children and our children's children. When we put wild systems first, we are passing on the gift of life to many species, including our own.

## Acknowledgments

I want to thank my agent, Charlotte Sheedy, who is smart, generous, and a bit of a miracle worker. I am grateful to Elisabeth Scharlatt, editor and publisher, for her care and goodwill, and to Andra Miller, who, along with Charlotte, first endorsed the idea of writing a book about seaweeds. To Chuck Adams for his perceptive edits, to my copy editor, Rachel Careau, whose superb eye gave this work the buffing it needed, to Brooke Csuka for her graceful support, and to Brunson Hoole for his skill and patience.

My son, Aran, has helped me right from the start with encouragement, editorial advice, and the belief that a story about seaweeds, and the shores where they grow, was worth doing. And thanks to my daughter, Caitlin, for her good help when I needed it most, her smart advice, and her many gifts; to Margot for her steady kindness and support; and to Dan for his gentle presence and caring heart.

This book is about the past and the present, but also, and perhaps especially, it points to the future. As I wrote it, my grandchildren were my compass. They will inherit, as will all our children and grandchildren, what we leave for them of the wild.

To Susan Jaques Curran for her steady friendship; to Susan Guilford, whose computer savvy saved me too many times to count. To Charles Guilford and Hugh Curran, who each brought me updates about seaweeds and seaweed harvests. To Norman and Paula Mrozicki and Mary and Steve Hildebrand, who made breaks from the long writing hours possible. A special thanks to Cynthia Thayer, and to Alan Furth and the people of the Cobscook Community Learning Center.

Intrepid first-draft readers Brendan Tierney, Claire Enterline, Patrick Tedesco, and Pam Chodosh guided me with insight and intelligence. I thank them for their smart advice. I learned more about coastal life and seaweeds from many people, including George Seaver, Liz Solet, William Kolodnicki, Raul Ugarte, Jessica Muhlin, Robin Alden, Glen Mittelhauser, Adam St. Gelais, Kara Ibarguen, Lauren Hall, and Mickey Scott.

My immense respect and thanks go to Brian Beal, Ted Ames, Paul Venno, Ken Ross, Susan Brawley, Claire Enterline, Nancy Sferra, Donna Kausen, Robin Hadlock

Seeley, Paul Molyneaux, Lindsay Tudor, Sarah Redmond, Seraphina Erhart, Shep and Linnette Erhart, Andrea DeFrancesco, Micah Woodcock, Brad Allen, Julie Keene, Larch Hanson, Carl Merrill, and Tollef Olson. Their clear voices and energetic spirits made this book.

A grant from the Alfred P. Sloan Foundation Public Understanding of Science and Technology Program helped to support my writing. For this I am deeply indebted to Eliza French and Doron Weber.

Thank you all.